POLYMERIC THERMOSETTING COMPOUNDS

COMPOUNDS

Innovative Aspects of
Their Formulation Technology

POLYMERIC THERMOSETTING COMPOUNDS

Innovative Aspects of Their Formulation Technology

Ralph D. Hermansen

Apple Academic Press Inc. Apple Academic Press Inc.
3333 Mistwell Crescent 9 Spinnaker Way
Oakville, ON L6L 0A2 Waretown, NJ 08758
Canada USA

©2017 by Apple Academic Press, Inc.

First issued in paperback 2021

Exclusive worldwide distribution by CRC Press, a member of Taylor & Francis Group
No claim to original U.S. Government works

ISBN 13: 978-1-77463-599-5 (pbk)
ISBN 13: 978-1-77188-314-6 (hbk)

Library and Archives Canada Cataloguing in Publication

Hermansen, Ralph D., author
Polymeric thermosetting compounds : innovative aspects of their formulation technology / Ralph D. Hermansen.

Includes bibliographical references and indexes.
Issued in print and electronic formats.
ISBN 978-1-77188-314-6 (hardcover).--ISBN 978-1-77188-315-3 (pdf)

1. Polymer engineering. 2. Polymers--Thermal properties. I. Title.

TP1087.H47 2016   668.9   C2016-903451-8   C2016-903452-6

Library of Congress Cataloging-in-Publication Data

Names: Hermansen, Ralph D.
Title: Polymeric thermosetting compounds : innovative aspects of their formulation technology / Ralph D. Hermansen.
Description: Toronto : Apple Academic Press, 2016. | Includes bibliographical references and index.
Identifiers: LCCN 2016021765 (print) | LCCN 2016024240 (ebook) | ISBN 9781771883146 (hardcover : alk. paper) | ISBN 9781771883153 (ebook)
Subjects: LCSH: Thermosetting plastics. | Thermosetting composites. | Heat resistant plastics. | Materials at high temperatures. | Materials at low temperatures. | Polymers. | Elastomers.
Classification: LCC TP1180.T55 H47 2016 (print) | LCC TP1180.T55 (ebook) | DDC 668.4/22--dc23
LC record available at https://lccn.loc.gov/2016021765

Apple Academic Press also publishes its books in a variety of electronic formats. Some content that appears in print may not be available in electronic format. For information about Apple Academic Press products, visit our website at **www.appleacademicpress.com** and the CRC Press website at **www.crc-press.com**

ABOUT THE AUTHOR

Ralph D. Hermansen

Ralph D. Hermansen has a BA/MS education, majoring in Chemistry from the University of New Mexico in Albuquerque. He also has 40 years' experience in materials science and engineering with an emphasis on formulating novel compounds from epoxies and polyurethanes. His experience includes working at several of the major aerospace companies and at two national laboratories. He retired from Hughes Aircraft Company, where he had functioned as a Senior Scientist. He and his team were sought out by design engineers in the various Hughes Aircraft divisions and by other companies to solve difficult custom-formulating problems. He has 21 patents resulting from these efforts.

He was active in SAMPE (Society for the Advancement of Materials and Processes Engineering), presenting several technical papers and serving as a Session Chairman. His first book, *Formulating Plastics and Elastomers by Computer*, was published in 1996. He also was an evening instructor within the Hughes Advanced Technical Education Program, teaching Materials Engineering and Pascal Programming Language for several years. Mr. Hermansen can be contacted for formulating advice through his website, CustomFormulator.com.

CONTENTS

LIST OF ABBREVIATIONS

AEC	Atomic Energy Commission
ANL	Argonne National Laboratories
BGE	butyl glycidyl ether
CPC	Chemical Products Corporation
CTE	coefficient of thermal expansion
CVCM	condensed volatile collectible material
DAP	diallyl phthalate
DBTDL	dibutyl tin dilaurate
ECA	electrically conductive adhesives
EDSG	Electro-Optics and Data System's Group
EPC	electronic power conditioner
FTIR	Fourier transform infrared spectroscopy
GSG	Ground Systems Group
HAC	Hughes Aircraft Company
HMS	Hughes material specification
HPLC	high-pressure liquid chromatography
HPS	Hughes Process Specification
IPDI	isophorone diisocyanate
MD	menthane diamine
MEK	methyl ethyl ketone
MEL	Manufacturing Engineering Laboratory
M & P engineers	materials and process engineers
MSG	Missile Systems Group
PBT	poly(butylene terephthalate)
PBW	printed wiring board
PU	polyurethane
RHMCC	reactive hot melt conformal coating
RSG	Radar Systems Group
SAMPE	Society for the Advancement of Materials and Processes Engineering
SBRC	Santa Barbara Research Company
SCG	Space and Communication-group

SJLE	solder joint lead encapsulant
TFNP	teflon-coated, non-porous
TMA	thermogravimetric analyzer
TML	total mass loss
TSD	technology support division
VCM	volatile condensable matter
WVR	water vapor recovered

PREFACE

Thermosetting plastics and elastomers, such as epoxies and polyurethanes, are very versatile materials. Due to the wide variety of ingredients available from which to formulate them, it is possible to create unique compounds to serve as adhesives, coatings, encapsulants, sealants, and a host of other applications. The author discusses the techniques of formulating polymeric compounds so that they will have a given set of properties. Properties could include mechanical properties such as ultimate tensile strength and elongation, or an electrical property such as volume resistivity. Perhaps, the material must be stable for years at 100°C or remain rubber like at −50°C. The greater the number of property requirements demanded of a compound, the more difficult is the task. In fact, the sought compound often does not exist and we will have to create it.

Several case histories are examined, in which the set of required properties for a new polymeric compound seemed extremely difficult to meet. Most of these problem-solving case histories were from aerospace electronics packaging needs. The author shares his thinking about how to approach these problems, drawing on a knowledge of different polymers and how these behave. The solutions to the problems were not obvious to someone skilled in the arts, which is the criterion for a patent. United States Patents were granted to the author and his associates for the novel solutions. The technologies developed have application beyond the specific applications cited. Potential spin-offs are discussed.

This book would be of interest to anyone who works with adhesives, encapsulants, coatings, and other polymeric compounds. It would be of great value to formulators of these compounds as well as those seeking a compound, which must have a certain combination of properties to function properly in an application. The book would also be of interest to those seeking a rewarding career in materials and process engineering. The author recounts his experiences in a national laboratory assisting in the fabrication of an atom smasher, in a major airplane company assisting design engineers, in an aerospace supplier company creating new compounds to fabricate windshields and canopies, in a chemical company custom-formulating flexible compounds for commercial manufacturing, in a telephone company formulating encapsulants for buried telephone cable splicing, and finally in a major aerospace company where he formulated custom materials for electronic assembly in satellites, missiles, military aircraft, and automobiles. A lifelong career is shared with the reader along with the secrets of successful formulating that he acquired along the way.

ACKNOWLEDGMENTS

I would like to thank Janet Hermansen, my very understanding and very supportive wife of 36 years. She made many sacrifices to allow me to focus full time on the research and writing the book.

Steve Lau, my partner in all these inventions, whose remarkable laboratory skills and extraordinary dedication made these inventions possible. Most patents cited in the book would never have happened without Steve's involvement.

Sandra Sickels, my editor at Apple Academic Press, whose patience and guidance were vital to the book's evolution.

Dr. Brent Pautler, who utilized ACD Labs ChemSketch software to make the chemical structures appearing throughout the book.

Kurt Hermansen, my son and legal counsel, whose advice has been essential. And his investigator, Maria Marcucilli, who helped me find old colleagues important to the narrative.

Gary Thunell, my son-in-law, whose computer skills and trademark/copywrite knowledge made writing the book feasible in the first place.

My friends and colleagues, Rod Patterson, Sidney Goodman, Phillip Magallanes, and Dr. Tom Sutherland, whose assistance throughout the writing has been invaluable.

Mark McGuire, my friend, neighbor and computer genius, who has never failed to help me with my personal computer problems.

INTRODUCTION

This book is about the art and science of custom formulating. Specifically, it is concerned with custom formulating of polymeric compounds, such as adhesives, sealants, casting compounds, coatings, encapsulants, and others. Formulating is the process of selecting polymeric ingredients and other additives into a recipe, which yields a useful compound. A typical formulation lists the ingredients, their amount, and any special processing needed. There are some similarities between polymeric formulating and creating a food recipe. Ingredients are added in a certain order and in a certain amount. Stirring or heating or other processing is done with the polymeric ingredients as it is done with the food recipe.

Instead of milk, flour, or shortening, polymeric formulations contain resins, curatives, mineral fillers, and other chemicals. There is a whole industry of ingredient suppliers, many having special properties sought by the formulator. Different polymers provide different properties. The polymer types normally associated with these applications are epoxies, polyurethanes, polysulfides, polyesters, and silicone rubbers. Hundreds of different ingredients are available for these polymers.

Custom formulating involves developing a new compound which satisfies a list of property requirements. Such a list is the major part of a material specification. The requirements generally come from a design engineer, who needs a specific material. Before custom formulating is ever considered, it has been determined that no such material currently exists. Custom formulating can be an expensive process and usually success cannot be guaranteed. The manufacturing department may also have input to the material specification. They may require that the material is safe to work with, will process in their existing equipment, and many other concerns.

There are literally hundreds of properties that might be important to a design engineer. These properties can be grouped under general headings such as physical properties (e.g., density, melting point, transparency, etc.), mechanical properties (e.g., tensile strength, shear modulus, Poisons ratio, etc.), electrical properties (e.g., dielectric strength, volume resistivity, dielectric constant, etc.), chemical properties (e.g., solvent resistance, outgassing, compatibility with cleaning agents), thermal properties (e.g., heat stability at some elevated temperature, glass transition temperature, etc.), and processing properties (e.g., pot life, cure temperature, cure time, etc.). The custom formulator has to develop a single compound able to satisfy several different property requirements.

The chapters in this book deal with actual custom-formulating projects from real applications. At times, some of the requirements seemed impossible to meet, especially in a single compound. The reader sees firsthand how the author and his lab partner dealt with these difficult problems. This book is valuable to those seeking a custom material because you cannot learn to custom formulate by taking college courses or attending a trade school. The author learned from decades of actual experience and he shares his knowledge and experience with you in this book.

PART I
Custom Formulating

CHAPTER 1

MATERIALS AND PROCESS ENGINEERING

CONTENTS

In the broader sense, this book is a story of my lifelong career in the field of materials technology. However, in a more specific sense, this book focuses on a particular productive period of my late career, when new polymeric materials were developed to solve specific engineering problems, and the solutions to those problems were inventive, that is, the solutions were not obvious to someone skilled in the art. Thus, the solutions were deemed worthy of U.S. patents. The technology behind such new material development is revealed and suggestions for new applications are discussed.

1.1   OVERVIEW OF THE PROFESSION

When people asked me what I did for a living over the years, I have answered the question differently depending on my employer at the time. I had job assignments that spanned the spectrum from manufacturing support to design support to applied research. I have also been titled as: research technician, engineer, manager or technical director, and senior scientist. The title that probably embraces all of these is "materials and process engineer" (also known as "M & P engineer"). The aerospace and defense industries employ large rooms filled with them. They typically specialize in specific fields like: metallurgy, ceramics, lubricants, composite, adhesives, and others. I started in the area of plastics and remained in it throughout my career.

1.1.1   IT IS A MULTI-DISCIPLINARY FIELD

Although a chemist by education, I have been attracted more to multi-disciplinary-type jobs than to jobs as a lab chemist. Working in the field of materials technology is definitely a multi-disciplinary work. An M & P engineer needs to understand the applications, where a material is to be used as well as how the material is processed in the manufacturing application. He/she thinks in terms of material properties, which are important to different applications. Sometimes the mechanical properties are very important, so knowledge about strength of materials is vital. Other times, it is the electrical properties or chemical resistance, which dominates one's interest. Or there may be restraints on how badly the material may contaminate its surroundings. For example, space satellites have optical surfaces, which must remain pristine. Materials used in satellites must not outgas as defined in a NASA requirement. How the material is processed might be a major concern. Both aircraft and space satellites are made of lightweight materials, which are structurally bonded together. However, the cure temperature for the adhesives is usually different. Aircraft structures are typically bonded together in autoclaves at temperatures above 250°F. Satellite parts are typically bonded at lower temperatures and the main assembly bonding is done at room temperature. The two processes are vastly different.

1.1.2 DIFFERENT KINDS OF M & P ENGINEERS

Manufacturing process engineers are a kind of M & P engineers, who are focused more on how to process a given material in an optimum manner than they are on the choice of the material. Usually, the design engineers have already dictated what materials will be used, so it is up to the process engineer to optimize the process and work with manufacturing to implement the process.

Some M & P engineers mainly focus on assisting design engineers in picking the right material for the application. They attempt to minimize the number of approved materials and for new applications, and also attempt to recommend one of the standard materials already being used. These M & P engineers are not highly interested in the formulation of the materials. If a material has been accepted into the company's approved materials list, then they deem it acceptable for new applications.

Finally, there is a type of M & P engineer who develops new polymeric compounds for the companies that market them. For specialized high-technology applications, these materials tend to be thermosetting plastics (also known as thermosets) rather than the thermoplastics. The difference between these two categories of materials is that thermoplastics are fully reacted linear polymers, which can be re-melted and molded into a new configuration. Thermosetting plastics, on the other hand, cannot be re-melted once they have reacted (i.e., cured). Thermosets are commonly sold as two-component systems, which can be combined and which then commence to cure into a three-dimensional polymer network. Epoxies, polyurethanes, silicones, and polysulphides are typical thermosetting plastics. This latter category of polymers is offered in the marketplace as compounds designed for specific applications. These compounds may be known as adhesives, sealants, encapsulants, potting compounds, coatings, and others.

1.2   THE WORLD OF FORMULATORS

The process of creating such compounds to satisfy specific applications and specialized properties is known as formulating. It is analogous to preparing a new recipe in the kitchen. Like the recipe, the formulation will consist of a list of ingredients, their proportion in the formulation, and any special processing that needs to occur. For example, ingredients may be mixed together under vacuum to assure that no air is mixed in. When companies, which manufacture such formulated compounds, prepare their product for sale; the process of combining these ingredients and packaging them is known as compounding. Although the technologists who develop formulated products have to know as much as M & P engineers, we usually refer them as formulators.

Although I have worked much of my career wearing most of those different M & P engineering hats, I have had more experience wearing the formulator's hat than any other. That formulating knowledge and experience was key to winning those patent awards, which will be discussed later in this book. Thus, I would like to tell you a little about how I gained that formulating experience.

1.2.1   HOW I BECAME A FORMULATOR

My plastics career started in 1959 and finally ended in 1998, a period spanning nearly 40 years. My first job in the plastics field was as a research technician at Argonne National Laboratory, near Chicago. The work involved preparing and testing epoxy resin formulations. I worked for a physicist, who was using plastics in support of the design of an atom smasher. I was also taking evening courses in an attempt to finish my Chemistry degree.

In 1964, I did get the Chemistry degree. However, it had involved moving to Albuquerque, New Mexico where the travel time was not tripled due to winter snow. I really loved the Sandia Mountains, the semidesert climate, and the academic atmosphere of the University of New Mexico.

I worked full time on the evening shift as quality assurance technician to support my growing family, but I missed the challenge of formulating.

Upon graduating, I hopped from Albuquerque to Seattle and back to Albuquerque as I learned to become a Materials and Process Engineer. I was becoming a foamed plastics specialist, first at Boeing Airplane Company (Seattle) where I supported the design engineers on the new 747 and 737 airplanes. Then I worked on weapon system support at Sandia National Laboratory. I did get to formulate a little, but it was not the main activity. I also continued part time with my education pursuing an M.S. in Chemistry at UNM.

1.2.2 THE FORMULATOR EMERGES

Although I did not plan it that way, from 1970 onward I primarily became a formulator. Secondly, I also found myself in a middle manager or team leader position in different jobs. Thirdly, my specialization in foamed plastics was over and I got exposure to a wider array of polymeric materials.

In late 1970s, my family and I were living in the Los Angeles. I was into formulating full time as Technical Director at Chemical Products Corporation, Western Division. My assignment was to build up a small development department to custom formulate urethane elastomers and vinyl plastisols for commercial applications. This was a growth experience for me in two ways: (1) I was switching completely out of foamed plastics and into the field of elastomers, and (2) I was in a small division, working directly for the general manager and with the operations manager, sales and marketing manager, and directly interfacing with customers. Gaining this top–down vantage was an advantage for future positions.

My formulating knowledge was expanded into the strange world of elastomeric polymers, where entropy plays a dominant role in mechanical behavior. The practical aspects of rheology were also part of my education. Chemical Products Company predominately made two-phase compounds; vinyl chloride particles in a liquid plasticizer for the plastisols and powdered mineral fillers in their polyurethane elastomers. The non-Newtonian behavior of these systems could cause pumps to jam and other problems. Two-phase systems also enable you to learn about packing theory. In other words, how by utilizing different particle sizes in the solid phase, you could change viscosity and flow characteristics.

Four years later, I was in the same area but now working for Sierracin/Sylmar Corporation, which designed and fabricated airplane windshields and canopies. I was hired as a senior engineer in their materials and processes department. I had not worked much with transparent plastics and elastomers previously, so I relished learning about this interesting new area. I was introduced to the field of transparent plastics and the property requirements of military windshields and canopies. I would spend the next 3 years pioneering the development of non-yellowing, water-white, rigid polyurethane plastics and get my first patent. Part V of the book tells the story.

In 1997, I was Manager of Chemical Research and Development for Communications Technology Corporation in Santa Monica. I took over group of formulators specializing in encapsulants for the telephone industry. Our customers were the telephone companies and the encapsulants had to be tested and qualified to formal specifications. This was really my first experience with formulating for electrical properties, such as volume resistivity, dielectric constant, and dielectric strength. Another new concept was re-enterability. Re-enterable encapsulants were weak enough to dig out chunks with your fingers. These concepts would be invaluable to me in my final job with Hughes Aircraft Company.

1.2.3　THE INVENTIVE PERIOD

I started working for Hughes Aircraft Company (HAC) on February 1, 1980. Hughes was divided into Product Groups and I was hired into the Space and Communications Group. We designed and built satellites. I was a Section head in the Manufacturing Engineering Laboratory (MEL). My section covered structural assembly, whereas my sister section covered electronic assembly. I had about a dozen engineers reporting to me. Half of them were concerned with composite parts' fabrication. The other half of my section covered diverse activities, such as adhesive bonding, honeycomb potting with syntactic foam, solar cell bonding, wire harness assembly, etc. The central theme of the structural assembly operations was to fabricate the structural parts of the satellites with the lightest possible structures, so that more electronics could be carried. The electronics part of the satellite was the real payload.

My inventive period actually started after I was transferred to the Technology Support Division (TSD) in the Electro-Optics and Data System's Group (EDSG). TSD had been the core technology center of the younger smaller HAC. I felt TSD was a better fit for my interests than MEL. It would be my home for the rest of my career. In fact, when I hit my magic 75 (age plus years of service) and could retire profitably, I continued as a contract engineer for another 2 years essentially doing the same job.

In my previous position, I could only work on structural assembly problems within SCG. Now I could offer my services to that area of SCG as well as the entire electronics area. Moreover, I could work on problems in the Radar Systems Group (RSG), the Missile Systems Group (MSG), the Ground Systems Group (GSG), Santa Barbara Research Company (SBRC), and any other part of the Corporation. At the time when I was transferred to TSD, HAC had approximately 80,000 employees. It was a huge corporation, with facilities scattered across Southern California and other states.

Between 1985 and 1998 when I retired, I acquired 20 of the 21 patents. They all resulted from finding a custom-formulated compound, which solved someone's engineering problem. Parts II through V of the book tells the case histories.

KEYWORDS

- materials and process engineer
- formulator
- space satellite

CHAPTER 2

THE ART AND SCIENCE OF FORMULATING

CONTENTS

Before jumping immediately into my real-life formulating adventures, I want to spend a little time acquainting the reader with the background to it. The following sections of Part I of the book discuss when a custom formulating effort is economically justified, how we go about defining our objectives, how to approach the problem of satisfying several property requirements, how to select a polymer family, and other considerations.

2.1 COST CONSIDERATIONS

We are trying to solve the problem of finding a material, which meets a unique set of requirements needed in a new application. If such a material actually exists out in the marketplace, we can save a great deal of money and time by finding it. The first step is to create a list of existing compound suppliers and then see if their catalogs have a matching candidate. For example, if we are searching for adhesives, we should start the search by a compiling a list of adhesive suppliers. Compound suppliers usually provide property data on their products, but their data sheet may not contain all the property data that you need to determine that they have a conforming product. The next step would be to talk with a technical representative of the company and determine if they have the information that you need. It may be necessary to obtain a sample of the product and conduct the necessary property test yourself.

Let's say that despite your best effort, no conforming material could be found. In the commercial world where thousands of pounds of required material might be purchased, there are supplier companies eager to develop such a material at their own costs. In the defense and aerospace industries, and other high-tech industries, often the quantities of material to be used are small, and the requirements can be challenging. It is unlikely that you will find a company with the skills to develop the compound that would do it for free. Some of them may be persuaded to do it under contract. Most of the companies possessing laboratories and competent formulators regard their lab time is very precious. They look at it from the standpoint, that of the various projects that they could be working on, which one will bring the biggest return for the time and money invested.

You may decide to try to develop the sought-after compound internally. You have the lab equipment and the formulating skills, so an internal development program may be the wisest course. You are going to be the project manager. So let me give you a word of caution. If you fail, the requesters are going to be very disappointed. There is an old saying, once burnt, twice shy. You may personally suffer from the failure, particularly if a great deal of money and schedule time is lost. It is important to honestly report the status of the project at intervals frequent enough that the requesters can cancel it if insufficient progress is not being made.

2.2 DEFINING THE OBJECTIVE

Before getting into the specific projects, I want to discuss some general points concerning new materials development. We learn best from our mistakes and I had become a big believer in spending sufficient time to get the objectives of a new project accurately defined. A common mistake made by young engineers is to run off to the lab with the objectives ill defined. Perhaps, they think their enthusiasm to get started quickly will impress their boss. Often the customer does not know him or herself, what is totally needed, what some of the properties are called, how to test for it, or what is most important to success. How the compound is applied, cure times and temperatures, skill level of the applicator, mechanical properties, electrical properties, chemical compatibility with other hardware, operating temperature range, humidity range, and many hidden requirements must be determined. I have learned that helping the requester through the process pays dividends. I remain skeptical that I have the full picture and keep my eyes and ears open to learn more. It can be embarrassing to boast that you solved the problem, only to learn your solution fails to meet a requirement no one told you about, but you were expected to automatically know.

The next step is to prioritize the list of objectives. We sometimes see requirements fighting each other as we try to meet them. Some properties are expensive to test for, and budget considerations may dictate that we save them for last. The greater the number of requirements we try to meet, the less likely that we will find an answer. If the development project is going to fail, it is best that we learn it as soon as possible. So, I look to see which requirements are most difficult to satisfy. Sometimes, on very challenging projects, I exclusively research formulations, which can satisfy the hardest requirement. If successful, attempt to satisfy the two most difficult requirements in one formulation and so on. If the minor objectives have to be compromised, the project may still be a success.

2.3   EXPERIMENTAL APPROACH

One way to assure that your development budget is not all spent fruitlessly is by making a list of the property requirement arranged from most difficult to meet to least difficult to meet. Concentrate on finding formulations, which can pass the hardest requirement. If you cannot accomplish this first step, perhaps the project needs to be either cancelled or redefined to something that can be done.

If you are successful meeting the hardest requirement, test those candidates to see if they also meet the second hardest requirement. If not, the next experimental effort should be to attempt to meet the two most difficult requirements in a single formulation. Again, if this proves to be too difficult, the direction of the project should be discussed with the customer. The process continues until all requirements are met or the project objectives redefined.

2.4   POLYMER SELECTION

2.4.1   EPOXIES AND POLYURETHANES

Most of the applications discussed in this book involve electronic packaging. This includes adhesives for holding components in place, adhesives for thermal transfer, conformal coatings for protecting components from corrosive agents, and conductive particles, which might cause a short circuit. It includes encapsulants and potting compounds to protect the components from mechanical and vibration forces as well as corrosive agents. Historically, two families of polymers have been used to formulate these products. They are polyurethanes and epoxy resin systems. Both polymer families offer the formulator great versatility in attaining customized properties. In general, epoxy resin systems cure to become hard plastics, whereas polyurethanes cure to become rubbery materials. As you will learn in this book, there can be exceptions to this general rule.

2.4.2   DEGREES OF RUBBER HARDNESS

Ninety percent of the inventions described in this book are rubbery materials. If you were to ask how soft or firm is the rubber being discussed, people in the trade usually give you a Shore A durometer number. The following table should help you translate those numbers into something familiar to most folks:

TABLE 2.1   Examples of Common Shore A Materials

Shore A durometer	Example
10	Synthetic fishing worm
20	Molded snakes and lizards
30	Art gum eraser
40	Pink pearl eraser
50	Inner tube rubber
60	Tire tread
70	"O" ring or ruby eraser
80	Typewriter eraser
90	Typewriter roller

Both polymer families become plastics or rubbers due to a chemical reaction between two or more ingredients in the formulation. We call that reaction, "curing." This curing most often results in a three-dimensional polymeric structure in the cured material. The frequency of the cross-links and the stiffness or flexibility of the polymer chain segments is controlled by the selection of ingredients. Let us consider what ingredient choices are available in the next two chapters.

2.5   FUNDAMENTALS OF POLYURETHANE FORMULATING

2.5.1   GENERAL

Polyurethanes are used in very large volume industrially as rigid foams, flexible foams, and coatings. They are used in smaller volume as thermoplastic resins, elastomers, and rigid plastics. Rigid foams was my area of expertise, when I worked at Sandia National Labs and Boeing Airplane Company. I had exposure to flexible foams at Boeing too. However, we are not very interested in foams for the purposes of this book. We are going to focus on elastomers and rigid polyurethanes, which are not really two distinct entities but rather a spectrum of polymers, having increasing hardness using the same family of ingredients. In fact, one of the ingredients, an isocyanate, is responsible for increasing the hardness of the members of this continuous series. The greater the isocyanate percentage, the harder the resultant material will be. The other ingredient, a polyol, usually has flexible polymeric chains between the reactive ends of the molecule. So the greater the polyol percentage in the formulation, the softer the resultant material will be.

2.5.2 POLYURETHANE POLYMERIZATION

Polyurethanes polymers result from the chemical reaction between polyols and polyisocyanates. Figure 2.1 shows the linkage reaction:

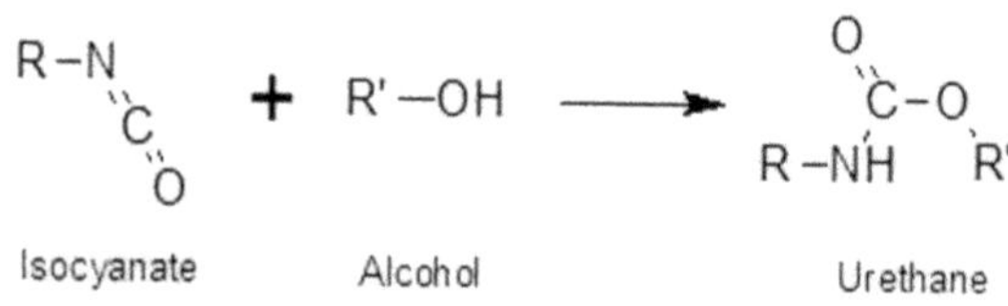

Isocyanate   Alcohol   Urethane

FIGURE 2.1   The isocyanate and alcohol reaction.

If this kind of chemistry is new to you, you may be wondering how do we build a polyurethane polymer out of isocyanates and polyols? Moreover, why would it be a three-dimensional polymer? The answer is that the reactive ingredients are at least difunctional and often trifunctional or more. Functionality is the number of reactive groups in the ingredients' molecule. Isocyanate ingredients are diisocyanates or greater. In other words, there are at least two isocyanate groups on every molecule. The following figure illustrates a diisocyanate.

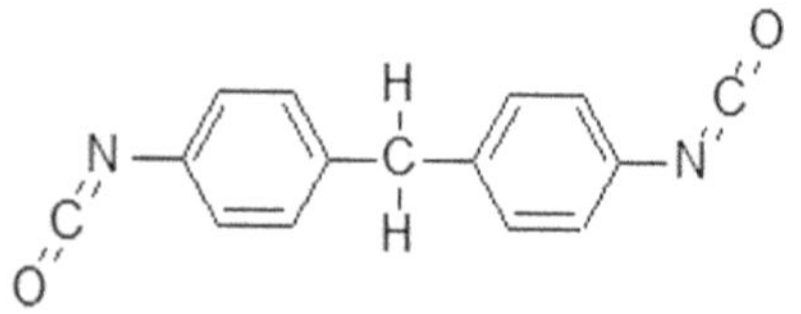

FIGURE 2.2   An example of a diisocyanate.

The situation is the same for the polyols. There are at least two hydroxyl groups per molecule (i.e., a diol). There could be three (i.e., a triol), four (i.e., a tetrol), and so forth. The following figure shows examples of a diol, triol, and tetrol where the hydroxyl groups are each on the terminal end of a polyoxypropylene chain segment.

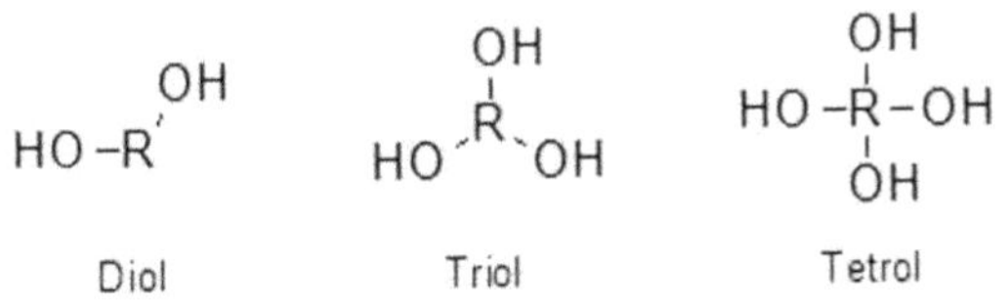

$$HO-R' \qquad HO \overset{OH}{\underset{R}{\diagup}} OH \qquad HO-\overset{OH}{\underset{OH}{R}}-OH$$

Diol   Triol   Tetrol

FIGURE 2.3   Diols, triols, and tetrols.

Consider the reaction product of diisocyanates and diols. If we assume there are no competing side reactions, we should expect very long linear polymer chains to form. We would have thermoplastic polyurethane. Such products are commercially available in the fully reacted state. They are usually elastomeric materials with excellent strength and wear resistance. Conveyer belts are usually made of polyurethane elastomers.

Now consider the reaction product of a diisocyanate and a triol. The polymer formed is a three-dimensional network polymer. Unlike the thermoplastic polyurethane, this one does not have a melting temperature, cannot be dissolved in compatible solvents, and wants to quickly return to its original shape when deformed. We can also predetermine how cross-linked our polymer will be by mixing diols and triols together in different proportions.

2.5.3   THREE-DIMENSIONAL POLYURETHANE POLYMERS

Did you play with Tinker Toys as a child? Formulating polyurethanes can be analogous to building a structure out of Tinker Toys. Just as the wooden dowel pieces came in a variety of lengths in the Tinker Toy box, so polyols also are commercially available in a variety of different molecular weights. If we select short-chain polyols for our formulation, we get stiffer, harder polyurethanes. Conversely, if we select long-chain polyols for our formulation, we get softer, more flexible polyurethanes. The reason for this has to do with the relative number of hard segments and soft segments in the polymer. The soft segments exist within the polyol ingredients. Predominately, these soft segments are polyether chains. They could also be polyester or polybutadiene chains. The hard segments of the polyurethane polymer form during the polymerization reaction. They are the urethane links formed by the reaction of isocyanate and terminal hydroxyl groups.

The Tinker Toy analogy can also be used to visualize what a three-dimensional polymer is like. Just as you build three-dimensional cars, buildings, and so forth with Tinker Toys, the polymer chemist builds three-dimensional polymers by allowing the polymer to grow in three dimensions.

Polyurethane compounds are often sold in a two-component kit form, where the isocyanate side is packaged separately from the polyol side. In order to form a polymer, we have to combine the two sides in the correct weight proportions and then thoroughly mix the batch to bring the different reactive molecules intimately together.

2.5.4 POLYURETHANE SIDE REACTIONS

Earlier we said that we would ignore side reactions when examining the basic chemistry. Now it is important that we know something of their existence.

2.5.4.1 THE ALLOPHANATE REACTION

One important side reaction is allophanate formation. Figure 2.4 shows the chemical reaction. As you can see, it is a cross-linking reaction. The allophanate linkage is formed by the reaction of a free isocyanate with the active hydrogen on the urethane group. The allophanate reaction is more likely to occur when there is an excess of isocyanate and cure temperatures are over 100°C.

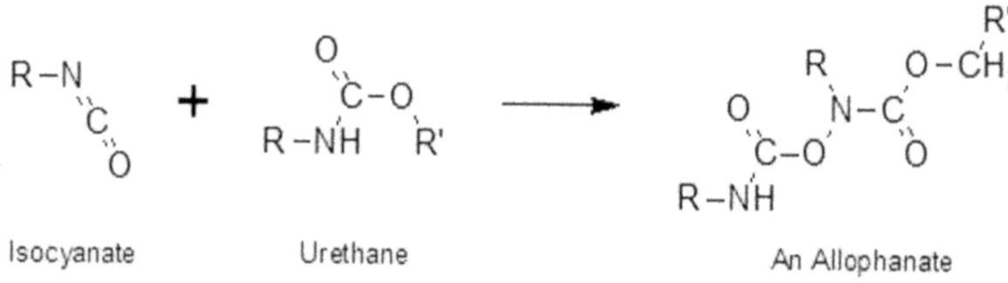

| Isocyanate | Urethane | An Allophanate |

FIGURE 2.4 The allophanate reaction.

THE WATER–ISOCYANATE REACTION

Polyols have an affinity for water, and despite special processing to remove water from the polyols, some water may still be present when the polyol and isocyanate components are combined. Those water molecules may undergo a reaction with the isocyanate groups. The chemical reaction is shown in Figure 2.5.

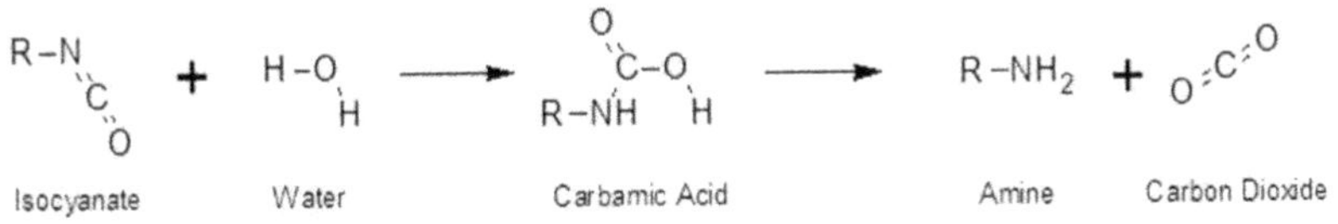

FIGURE 2.5   The water–isocyanate reaction.

The isocyanate group is converted to a primary amine group and carbon dioxide is formed. It escapes as a gas and may cause bubbles to form in the polyurethane product. For the polyurethane foam industry, this is a desirable thing and water may have been intentionally added to create the gas. However, if you are striving for a bubble-free product, this reaction may cause you anguish.

The primary amine is short lived. It immediately reacts with another isocyanate group and forms a urea group.

2.5.4.2   THE BIURET REACTION

If urea linkages exist in our polyurethane polymer, free isocyanates may react with them if conditions are right. The chemical reaction is shown in Figure 2.6.

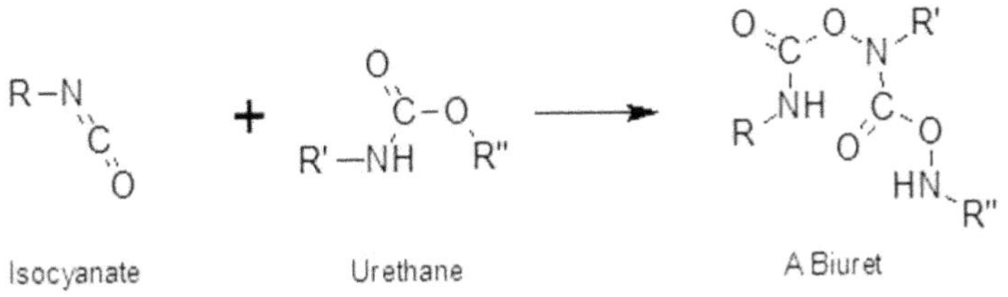

FIGURE 2.6   The biuret reaction.

The biuret formation is a cross-linking reaction.

2.6   FUNDAMENTALS OF EPOXY RESIN FORMULATING

2.6.1   *THE DIFFERENCES BETWEEN POLYURETHANES AND EPOXIES FROM A FORMULATOR'S VIEWPOINT*

Next, we shall discuss epoxy resin formulating. As we end our discussion of polyurethane formulating, I want to point out some of the important differences between these polymer families:

(1) The presence of water in polyurethane compounds is a big problem, if we do not want to make foam. The problem also affects the environment where we use the polyurethane compound. Some substrates, like wood, have water content that can cause foaming at those surfaces. The problem virtually disappears when we work with epoxies.

(2) The reaction rate of polyurethane compounds can be fine-tuned using catalysts. There are numerous different catalysts available. The catalyst that I used in the polyurethane compounds discussed in this book was dibutyltin dilaurate (DBTDL).

(3) The available epoxy resins and curing agents are very useful if you want to make a hard, somewhat brittle plastic. The available polyols and isocyanates are very useful if you want to make a rigid or flexible foam or a rubbery product. However, if you want to make a rubbery epoxy or a hard polyurethane plastic, you are more or less on your own. Strangely enough, I found myself needing to do both of those things.

2.6.2   *EPOXY RESIN/CURATIVE SYSTEMS*

The epoxy category of plastics is probably more familiar to folks than are polyurethanes. Hardware stores typically carry several different kinds of epoxy adhesives sold as two-component kits. You squirt out two equal size quantities of resin and hardener, stir them intimately, and apply the mixture to broken parts, and after a while, the parts are strongly bonded together.

Epoxy resins are capable of reacting with epoxy curatives to form a three-dimensional polymer. Figure 2.7 shows the chemical structure of a common epoxy resin. Figure 2.8 shows the structure of a diamine curative, capable of converting the epoxy resin into a plastic material.

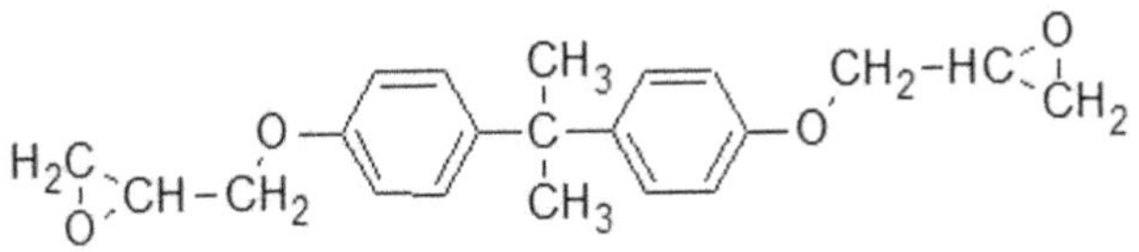

FIGURE 2.7   Example of an epoxy resin.

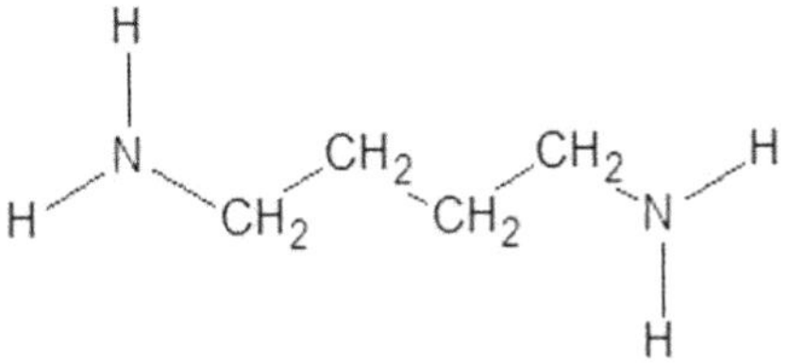

FIGURE 2.8   Example of a diamine curative.

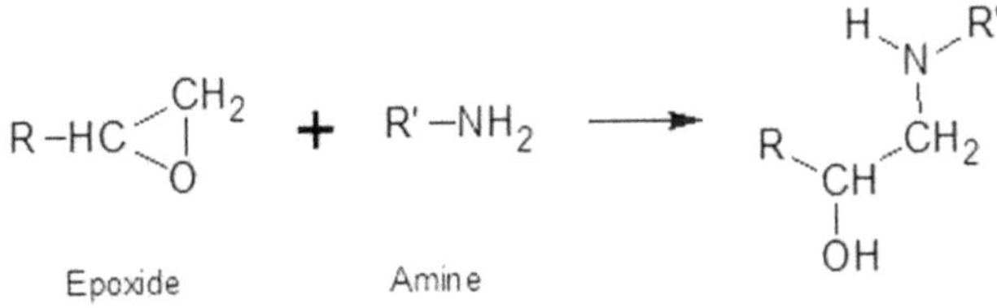

FIGURE 2.9   The epoxy–amine reaction.

A chemist would say the resin component is a diepoxide and the curative could be one of several different chemicals, although we know that it is difunctional for sure and more likely polyfunctional. One category of epoxy resin curatives are aliphatic polyamines. Amine groups are called primary, secondary, or tertiary depending upon how many hydrogen atoms are attached to the amine's nitrogen atom. Two hydrogen atoms and it is primary, one hydrogen and it is secondary, zero hydrogen atoms and it is tertiary. These hydrogens are called active hydrogens and each one of them can react with an epoxy group forming a linkage. Consider hexane diamine, one of the curatives we will discuss later. It has four active hydrogens, which makes it tetrafunctional. Hexane diamine is very reactive and can join four epoxy resin chains together, thus building a cross-linked polymer. In general, aliphatic polyamines can cure epoxy resins at room temperature. Aromatic polyamines are less reactive and can only partially cure the epoxy resin at room temperature. The reaction stops at what is called a "B" stage. The cure can be completed later in time by using an oven cure to drive the reactions to completion.

Tertiary amines have no active hydrogen atoms, but when mixed with epoxy resins are capable of catalyzing a polymerization reaction between epoxy resins. It is called homopolymerization.

2.7 EFFECTIVE EXPERIMENTAL TECHNIQUES IN FORMULATING

2.7.1 MASTERBATCH BLENDING

We have two different formulations, X and Y, each having one or more of the properties needed in the final product. Assuming that the ingredients of the two formulations are compatible with each other, it can be very productive to make intermediate blends of compounds X and Y to determine how the properties of interest change as the formulation changes.

TABLE 2.2   Experimental Plan for Blending Masterbatches

	Trial 1	Trial 2	Trial 3	Trial 4	Trial 5
Compound X	100%	75%	50%	25%	0%
Compound Y	0%	25%	50%	75%	100%

Table 2.2 shows how five trial formulations can be prepared by blended masterbatches X and Y. The same five trial formulations could be prepared by weighing out each of the ingredients in X and Y for each of the trials. Blending masterbatches is a better way to prepare them because weighing errors are reduced. If we were to get erratic property results instead of a smooth curve, our first suspicion would be weighing errors were made.

There are an infinite number of formulations intermediate between formulae X and Y. However by testing X, Y, and these three intermediates, we have a very good idea what the properties of these infinite number of intermediates are.

2.8 MAKING PARTIALLY REACTED INTERMEDIATES (PREPOLYMERS)

Sometimes the key to attaining a formulating objective is through the formation of a prepolymer. What is a prepolymer? It is a material in a stage somewhere between raw ingredients and a fully cured state. For example, we have an epoxy resin and a diamine curative. We could mix them together as is and get a cured epoxy plastic. That is called the one-shot method. However, suppose the resin and curative were incompatible and separated like oil and water after being combined, it would appear at first glance that we cannot use them together. However, there is a way to make them more compatible. We could make an amine-terminated prepolymer or an epoxy-terminated prepolymer from the ingredients and improve their compatibility.

How do we make a prepolymer? First, we have to know something about the functionality of the epoxy resin and the epoxy curative. If the functionality is too high, we will get cured plastics instead of a liquid prepolymer. Difunctional reactants are safe, and sometimes trifunctional reactants are also safe. Let's say we had a difunctional epoxy resin and a tetrafunctional diamine. The four difunctional epoxy resins molecules could react with one diamine molecule to form a tetrafunctional epoxy-terminated prepolymer. This prepolymer could then be fully reacted with diamine curative in the ratio of one prepolymer molecule to one diamine molecule. There can be other reasons for making prepolymers besides compatibility. For example, a prepolymer will reduce peak exotherm temperature compared with the one-shot approach.

Polyurethanes can also utilize the prepolymer concept. There is often incompatibility between polyisocyanates and certain polyols. Isocyanate-terminated prepolymers abound in the industry for this reason. Sometimes the reason is not compatibility, but toxicity. TDI is the standard abbreviation for tolylene diisocyanate. This is a very disagreeable chemical in its pure form, but is far more acceptable in its prepolymer form.

KEYWORDS

- **formulating**
- **side reaction**
- **polyurethane**
- **epoxy resin**

PART II
Thermal Transfer Adhesives
for Space Electronics

CHAPTER 3

THERMAL TRANSFER, FILLETING ADHESIVES

CONTENTS

Space electronics are unique because they operate in an airless environment where convection cooling is not possible, yet the heat from components must be dissipated to assure their reliable operation for 10 years or more. Satellites in high orbits cannot be visited and repaired. In this section, we shall discuss the engineering problems, formulating efforts, and qualification of new compounds solving those problems. Each of these case histories resulted in a U.S. patent.

3.1   THE POLYSULFIDE REPLACEMENT PROJECT

In the fall of 1985, I transferred within Hughes Aircraft Company from the Manufacturing Engineering division, in the Space and Communication group (SCG), over to the Technology Support Division (TSD) within the Electro-Optical and Data Systems Group (EDSG). At SCG, my range of authorized problems was limited to spacecraft structures, whereas at TSD, any problem in the entire HAC was fair game as long as it was within my area of expertise and the group with the problem, provided funding for me to work on it.

The first big problem assigned to me actually came from SCG, my old group. However, it had nothing to do with my old domain, structural assembly. Instead, it had to do with electronic assemblies, namely, populated printed wiring boards. A mineral-filled polysulfide adhesive had been in use as a filleting adhesive around components and also in the form of a film adhesive under flat-bodied components. The purpose of these adhesives was to help conduct heat away from the components to give them longer service life. On Earth, we simply use a fan to blow air over the electronic components in order to cool them, but there is no air in space so instead, we use a thermally conductive adhesive to keep the components cool. The cooler that the electronic components can operate, the longer they will last. Considering the multimillion dollar costs of building a satellite and launching it into an orbit, it is essential that it operate for as long as possible.

One of the requirements that the materials to be used in a satellite must fulfill is that they pass the NASA outgassing requirement. Meeting this requirement assures that minimal organic vapors are released from spacecraft materials. These vapors could deposit on solar cells, mirrors, so forth and become opaque due to solar radiation. The polysulfide adhesive failed this NASA requirement very badly and consequently was no longer acceptable for use on many programs. My assignment was to emerge with a new adhesive, which had equal or better properties than the existing polysulfide adhesive, but definitely passing the NASA outgassing requirements. I was the principal investigator of a small team of engineers assigned to this effort..

3.2 TERMS AND DEFINITIONS

3.2.1 WHAT IS A FILLETING ADHESIVE?

The currently used polysulfide adhesive was purchased as a one-component, frozen premix adhesive. Once thawed, it had the consistency of a thixotropic paste, that is, it could be applied as a fillet around resistors, capacitors, diodes, transistors, and so forth using a spatula to shape the fillet and the fillet would hold its shape during the 200°F cure cycle. The thermally conductive adhesive was also used in a film version, but the filleting adhesive replacement was the most urgent part of the project because it was 85% of the total polysulfide adhesive usage.

3.2.2 WHAT IS THERMAL CONDUCTIVITY?

Thermal conductivity is the property of a material to conduct heat. Metals have high thermal conductivity, and glasses and ceramics have much lower thermal conductivity. Plastics have even lower thermal conductivity than glass. Air, or vacuum for that matter, has the lowest thermal conductivity. There is essentially no heat conduction and this is one of the biggest problems of spacecraft electronics. Heat cannot escape from the components. Application of an adhesive provides a path for the heat in the electronic components to dissipate. We can increase the thermal conductivity of an adhesive by adding a suitable filler to the polymeric matrix. However, the filler cannot be electrically conductive or even semiconductive because it would modify the performance of the electronic circuitry. Inorganic minerals are generally acceptable as fillers, and experience has shown that tabular alumina is one of the most effective fillers for enhancing thermal conduction.

Tabular alumina is the filler of choice for use in thermally conductive adhesives for electronic assembly because it has excellent dielectric properties and it has been densified by rapid sintering. The particle shape is also important. Tabular alumina has large, well-developed hexagonal tablet-shaped crystals. Thus, the particles in the adhesive have far more than just point contact with each other.

3.2.3   WHAT IS NASA OUTGASSING?

NASA requires the materials used to fabricate spacecraft to pass their non-outgassing requirements. Samples of the test materials are subjected to heat and vacuum as per the procedures of ASTM E595. This consists of subjecting the test samples to 24 h at 125°C in a vacuum of 5×10^{-5} torr. The samples are weighed before and after exposure. The requirements for passing the NASA outgassing test are: Total mass loss (TML) less water vapor recovered (WVR) must be equal to or less than 1.0%. Condensed volatile collectible material (CVCM) must be equal to or less than 0.1 %. One of the main reasons for the requirement is that volatile organic compounds might deposit on optical surfaces, such as solar cells. Solar radiation might turn the residues opaque rendering the solar cells inoperative. Moreover, any optical device on a satellite or spacecraft might be rendered inoperative by the same deposits.

3.2.4   WHAT IS A GLASS TRANSITION TEMPERATURE?

Polymeric materials are usually either rigid plastics or rubbery materials. We define those materials as either rigid plastic or rubber based on how they behave at room temperature. However, if you raise the temperature of a rigid plastic, at some point it will become rubbery. If you cool a rubbery material to a low enough temperature, it becomes a rigid plastic. The glass transition temperature is the temperature which is intermediate between the rigid and rubbery states. Some polymeric materials are in their glass transition zone at room temperature. They have a leathery behavior to them. Instead of being springy, they slowly come back from a deformed condition. The glass transition temperature is the temperature of maximum damping and one test method (i.e., torsion pendulum) uses this fact to determine the glass transition temperature.

For the thermal transfer adhesive application, we prefer the lowest possible glass transition temperature so that the adhesive is rubbery over the operating temperature range of the application. We desire a low modulus, rubbery state so that stress is minimized on the components during thermal cycling. Glass-bodied diodes are most sensitive to such stress. Rigid adhesives can crack or break these components. A rigid adhesive would cause fragile components to break due to differing coefficients of thermal expansion. Moreover, a slightly weak, rubbery material allows the removal of components from the circuit board without destroying them in the process.

3.2.5   WHAT IS A FROZEN PREMIX ADHESIVE?

Many of the adhesives used in aerospace applications are two-component systems. The components are weighed out according to the recommended mix ratio and thoroughly stirred together. A chemical reaction called polymerization then occurs, which converts the liquid mixture to either a plastic or rubbery material. Epoxies, polysulfides, polyurethanes, and other polymer types are commonly packaged in this manner.

Experience has shown that performing such weighing and mixing operations on the manufacturing floor is problematic. Weighing errors or under-mixing can lead to needless rework and even schedule delays. Undiscovered problems with the adhesive might affect the function of the hardware in space, where repair is not feasible. Thus, it has become standard practice to outsource the weighing, mixing, and packaging of two-component adhesives to a specialty house, where they become one-component, frozen premix adhesives. After quick freezing to stop the polymerization reaction, the cartridges are stored in freezers until needed. The specialty house may be required to sample the frozen premix cartridges and test for adhesive strength or other properties to confirm whether processing was done correctly.

The polysulfide thermally conductive adhesives have been obtained as frozen premix cartridges and the new adhesive, replacing them, was expected to be amenable to similar frozen premix packaging and storage.

3.3   SEEKING A REPLACEMENT FOR THE OUTGASSING POLYSULFIDE FILM ADHESIVE

A steering committee was formed to direct and monitor the progress in replacing polysulfide adhesives with non-outgassing equivalents. It consisted of members from three organizations, namely (1) SCG Digital

Electronics Engineering, (2) SCG Manufacturing, and (3) Technology Support Materials and Process Engineering.

3.3.1 LESSONS FROM THE 1983 VENDOR SEARCH

I obtained a final report on a previous unsuccessful attempt to replace the polysulfide adhesive ending in 1983. I carefully read this report searching for clues on how I might solve the problem this time. I learned that although none of their candidates was deemed acceptable, the two candidates, which came closest to being acceptable, were polyurethane elastomeric formulations. As expected, both candidates contained alumina powder as a filler.

The two promising candidates were rejected because (1) they were too strong cohesively to allow rework without damaging components, and (2) they also failed because their storage life as a frozen premix was unacceptably short. This information was very useful. It is relatively easy to modify an adhesive to make it weaker. So there may hope for these candidates yet. Extending their freezer shelf life might require some research.

3.3.2 WHAT PROPERTIES MUST THE NEW ADHESIVE HAVE?

Table 3.1 lists the properties, test methods, and requirements for the new thermally conductive filleting adhesive.

TABLE 3.1 Property Requirements for the Thermally Conductive Filleting Adhesive

Property	Test Method	Requirement
Outgassing	ASTM E595	TML $\leq 1.0\%$, CVCM $\leq 0.1\%$
Thermal conductivity	Calora conductometer	>0.35 BTU/h-ft-°F
Glass transition temperature	Thermomechanical analysis	−44°C or lower
Coefficient of thermal expansion	Thermomechanical analysis	17.1 cm/cm/°C max
Shore A durometer	ASTM 2240	55–80
Volume resistivity	ASTM D257 at 500 V	$>10^{12}$
Dielectric strength	ASTM D149	>350 V/mil

TABLE 3.1 *(Continued)*

Property	Test Method	Requirement
Dielectric constant	ASTM D150	@ 1 kHz, 6.0 max
		@ 10 kHz, 6.0 max
		@ 100 kHz, 6.0 max
		@ 1000 kHz, 6.0 max
Dissipation factor	ASTM D150	@ 10 kHz, 0.026 max
		@ 100 kHz, 0.026 max
		@1000 kHz, 0.026 max
Lap shear strength	ASTM D1002	At 75°F, 100-400 psi
(aluminum to aluminum)		At 200°F, 50-150 psi
		At −60°F, >100 psi
Lap shear strength	ASTM D1002	
(a) Nickel plate		100–400 psi
(b) Solder plate		100–400 psi
(c) Polyimide laminate		100–400 psi
Cleavage strength	ASTM D1062	50–150 piw
Flux removal solvent compatibility	Cyclic exposure to immersion in isopropanol and Freon TE vapors	No deterioration of adhesive or damage to bondline
	ASTM D471 and ASTM D1002	
Conformal coating compatibility	Visual examination of Uralane 5750 after application to filleted bonds	No deterioration of adhesive or coating. No effect on cure of coating
Work-life confirmation	Prepare lap shear specimens after 2 h aging. Test per ASTM D1002	100–400 psi
Thixotropy or sag resistance	Measure sag of bead on vertical surface after 1 h	1.3 cm max
Storage life at −40°F	Conduct work-life test after 3 months	Must pass
Cure to handling in 20 min at 200°F	Determine Shore A durometer hardness per ASTM D2240	Non-tacky texture, 30 Shore A min

3.3.3   COMPANIES SUBMITTING SAMPLES IN THE 1985–1986 EFFORT

Five different companies responded to our request to submit samples for our appraisal. Samples arrived toward the end of 1985. A quick screening test was used to determine if the samples would be worth advancing to the next, more expensive level of testing. The screening test program included: (1) the NASA outgassing test per ASTM E595, (2) examining the working life and cure at 200°F, and (3) a reworkability test, where we attempted to tear the cured sample by hand to appraise its ease of removal.

Three of the five companies were eliminated in the preliminary screening phase. Their submittals failed in a critical way. Failure to pass the NASA outgassing requirements was the major reason for dropping these candidates. The Crest Products Company and Furane Plastics Company were not rejected. Crest had submitted a flexible epoxy adhesive, which passed the NASA outgassing test. However, the cured sample had a tacky feel to it and the adhesive cured too slowly. Crest was encouraged to modify their submittal and resubmit. Furane Plastics Company submitted a polyurethane adhesive, but they had used aluminum powder instead of alumina powder as a filler. This was a misunderstanding and so they were encouraged to resubmit samples using the proper filler (i.e., alumina powder).

3.3.4   FURANE PLASTICS AND CREST PRODUCTS CONTINUE

Crest Products Company was unable to eliminate the tackiness or to get their candidate to cure faster. Furane Plastics Company submitted two new candidates which were alumina-filled polyurethanes. One of them was based on a polyester-based polyol ingredient. It failed the NASA outgassing requirement badly and was eliminated from further evaluation. The other polyurethane candidate was based on a polybutadiene polyol ingredient and it looked very promising.

At this point in the program, I sensed that this candidate (i.e., Uralane 88840) could be further modified to meet our criteria and thereafter, I personally monitored the Furane Plastics effort. This meant my making numerous trips to their facility and encouraging them to continue modifying the formulation until we had an acceptable product. My contact at Furane Plastics was Ed Clark. On my drive to work in El Segundo, it was easy for me to detour to their plant in Glendale. Sometimes, I was there before

Ed arrived himself. He told me that it was a new experience to have the customer arrive to work before him.

The most vitally important modifications that I suggested to him increased the storage life of the frozen premix version from 1 month to over 3 months. One month was unacceptable to SCG Manufacturing Department, and even 3 months was borderline acceptable to them. A critical suggestion was to replace their aromatic diisocyanate ingredient with an aliphatic diisocyanate ingredient. My previous experience had taught me that the change would reduce the reactivity of the polyol/isocyanate polymerization reaction modestly, but would reduce the reactivity of the water/isocyanate reaction considerably. Moisture must be excluded from the adhesive mixture as thoroughly as possible. Moisture must be removed from the polyols and fillers by heating them slightly above the boiling point of water under vacuum.

My suggestion proved to be successful. The storage life was adequately extended by the use of aliphatic diisocyanates. This was a major breakthrough! The whole category of polyurethane elastomers with their vast versatility was suddenly available because I could now make them into frozen premix, one-component systems. I was very happy!

The back of the problem had been broken and I was able to help Furane Plastics develop a product that exceeded our expectations for a filleting adhesive. One other important modification was the removal of a phthalate plasticizer from the furane formulation. This contaminant was analyzed by our chemistry laboratory during an analysis of CVCM from the outgassing test.

The final furane formulation was named Uralane 7760 adhesive. The extensive testing of Uralane 88840 and its modification to become Uralane 7760 was no small effort. It involved testing processing properties, mechanical properties, thermal properties, environmental properties, and electrical properties. The co-coordinator of this testing effort was Mark Weaver of TSD. He issued a report in May 1986 showing the new adhesive fully complied with all of the requirements.

3.3.5 OUR CONCURRENT IN-HOUSE FORMULATION DEVELOPMENT EFFORT

Knowing that polybutadiene-based polyurethanes had a good chance of meeting the requirements of the thermal transfer adhesive, we also had an in-house formulation effort underway. Initially, this effort was a safeguard against all of the submitted candidates being rejected as unacceptable and my team failing in the assignment. Our biggest drawback was the lack of proper processing equipment. We needed a sturdy mixer to combine the filler and polyols and we needed to do this mixing under vacuum at elevated temperatures. We knew that a Ross planetary mixer was perfect for the job, but many long months went by before we were able to obtain this equipment. Fortunately, Furane Plastics did have the proper equipment and the project was successful because of that. What we were learning in our lab was helping me guide Ed Clark and his team at Furane Plastics in a successful direction.

3.4   REFINEMENT INTO URALANE 7760 ADHESIVE AND QUALIFICATION TESTING

3.4.1   FINALIZED CANDIDATE—URALANE 7760

The final tweaking of the Uralane 7760 formulation was to examine two different alumina filler loadings: 60% and 70%. After examining the test data, we decided to go with the higher loading version. That final version was fully tested as per the requirements and it passed all the tests. The property of utmost importance was NASA outgassing. Uralane 7760 passed with flying colors. It had 0.30% total mass loss against a requirement of 1.0% max. It had A CVCM of 0% against a 0.1% max requirement. These data are especially welcome considering that polysulfide adhesive had a TML of 19%.

3.4.2   QUALIFICATION OF URALANE 7760

Although my focus in this chapter is to disclose the formulation development story of a new material, a whole team of people made the overall effort successful. Some were working on various aspects of qualification testing, others on actual production verification, and still others doing the writing of new specifications for this new material. Moreover, the overall effort was much larger than we have space for in this chapter. Multimillion dollar satellites were at stake. There is a great cautiousness in making changes to a proven system in the aerospace business. On the other hand, NASA was not willing to accept the polysulfide adhesive on many new programs.

Uralane 7760 had a glass transition temperature of $-63°C$ versus a requirement of $-44°C$ max, an excellent result. This new adhesive could be considered for usage in other applications requiring flexibility to very low temperatures. The polybutadiene polymer chains in the adhesive were responsible for attaining such a low T_g. Moreover, the electrical properties were excellent from a dielectric standpoint. The volume resistivity was two decades higher than the requirement, the dielectric strength was 600 V/mil, and the dielectric constant and dissipation factor were well within the requirement limits.

Three different batches of Uralane 7760 were obtained from Furane Plastics and tested for batch to batch consistency, which they passed. An accelerated space aging evaluation was conducted using (a) 5 days aging at 175°F and 10^{-5} torr, and (b) 4 days aging at 225°F and 10^{-5} torr. Shore A durometer, weight change, lap shear strength, volume resistivity, and dielectric strength were compared before and after aging. The results were all favorable.

3.4.3 NEW SPECIFICATIONS FOR THE NEW ADHESIVE

A Hughes materials specification (HMS 2289) and a processing specification (HPR 63032) for the new adhesive were prepared by members of our team. It was deemed necessary to add a sag resistance test to HMS 2289 because the previous polysulfide filleting adhesive was inconsistent from batch to batch and making the test an incoming requirement helped assure consistency with the new filleting adhesive. For the process specification, HPR 63032, cure times for different cure temperatures were determined for Uralane 7760 and added to the specification.

The new adhesive, Uralane 7760, was recommended for usage. Some programs directed that there should be immediate conversion to the new adhesive. I issued a final report on the qualification of Uralane 7760 as a replacement for the polysulfide filleting adhesive application on January 13, 1987. This major part of the polysulfide replacement project was highly successful. The film adhesive version was the remaining issue.

3.5   ADDITIONAL QUALIFICATION TESTING

Design engineers are conservative by nature and in the case of space electronics, they are ultra conservative. It is better to test for every conceivable thing that can go wrong, and then learn about if a multimillion dollar satellite fails to function. Additional tests were deemed necessary.

3.5.1   GLASS-BODIED DIODE COMPATIBILITY

Uralane 7760 adhesive would soon be adopted on all satellite programs as a filleting, thermal transfer adhesive. The question was raised about the compatibility of the new material with glass-bodied diodes. Prior to polysulfide filleting adhesive, glass-bodied diodes were sometimes broken during thermal cycling due to induced stresses. Although it is true that Uralane 7760 is also an elastomeric compound and should be acceptable for these sensitive components, a test is worth a thousand expert opinions.

The test consisted of mounting glass-bodied diodes on fiberglass boards or nylon boards using Uralane 7760 and polysulfide adhesive as a control. Then, the specimens are exposed to five thermal cycles between −55°C and 125°C. The diodes were inspected for breakage.

The results were deemed very favorable. Of the 56 glass diodes tested, only two were damaged. One adhered with polysulfide cracked where it had been intentionally scratched. The other adhered with Uralane 7760 was chipped on one edge. It might have occurred before thermal cycling. Uralane 7760 filleting adhesive was deemed safe for glass-bodied diodes.

3.5.2   SPACE AGING SIMULATION

3.5.2.1   TEST PROCEDURE

Although the NASA outgassing test provides information about how a material behaves in a space environment, it is only a 24-h exposure. The Steering Committee members, D.D. Miller and R. Kovnat, wanted more assurance that the new adhesive, Uralane 7760, would survive for 10 years or more in a space environment. To this end, a test program was devised: Specimens of both Uralane 7760 and the current polysulfide adhesive were prepared. We had a long history of how polysulfide behaved in orbit.

Two sets of test specimens were exposed to 7 days at 175°F and hard vacuum (about 10^{-5} torr). One set of specimens was tested after this exposure, while the second set was further exposed to 4 days at 225°F and hard vacuum. These specimens were then also tested. Specimens with no aging were also tested to establish the initial condition.

3.5.2.2   TEST RESULTS

Table 3.2 below summarizes the comparison of Uralane 7760 adhesive and the filleting adhesive, which it replaces. Notice the severe deterioration of polysulfide after the 225°F plus hard vacuum exposure.

TABLE 3.2   Comparison of Uralane 7760 and Polysulfide Adhesives after Heat Plus Hard Vacuum Aging

Property and test method	Uralane 7760 adhesive	Polysulfide adhesive
Shore A durometer Per ASTM D2240	Initial condition: 73–75 After 7 days at 175°F: 74–75 After 4 days at 225°F: 79	Initial condition: 60–61 After 7 days at 175°F: 65–68 After 4 days at 225°F: 82–84
Weight and dimensions	Minor change	After 7 days at 175°F: 5% weight change After 4 days at 225°F: 19–20% weight change
Lap shear strength, psi Aluminum-to-aluminum, per ASTM D1002	Initial condition: 380 After 7 days at 175°F: 380 After 4 days at 225°F: 510	Initial condition: 290 After 7 days at 175°F: 160 After 4 days at 225°F: 160
Volume resistivity, ohm-cm Per ASTM D257	Initial condition: 1.86×10^{15} After 7 days at 175°F: 2.79×10^{15} After 4 days at 225°F: 1.64×10^{15}	Initial condition: 1.16×10^{13} After 7 days at 175°F: 8.99×10^{14} After 4 days at 225°F: too damaged to test
Dielectric Strength, V/mil, per ASTM D149	Initial condition: 828 After 7 days at 175°F: 803 After 4 days at 225°F: 773	Initial condition: 531 After 7 days at 175°F: 374 After 4 days at 225°F: too damaged to test
Rheometrics	Minor change	Major increase in dynamic shear modulus at 50°C
Visual appearance	Very slight darkening with increased temperature exposure	Major embrittlement and fragmentation at 225°F/vacuum

Uralane 7760 is far superior to polysulfide adhesive in a simulated space environment.

3.6 OUR SAMPE PAPER AND PRESENTATION

My department manager, David Chow, encouraged me to present our accomplishments at an upcoming SAMPE conference. SAMPE is short for the Society for the Advancement of Materials and Process Engineering. David and I co-authored a paper entitled "Novel Thermally-Conductive, Elastomeric Adhesive for Electronic Component Assembly," which was accepted by SAMPE for publication. I presented the paper in mid-June 1988 at the SAMPE Electronic Materials Conference in Seattle. My wife, Janet, joined me on the trip. It was a treat to visit Seattle for both of us. One experience was especially enjoyable. We, along with several of the conference attendees, took a motor launch from Seattle Harbor to Blake Island for a native American show, a clam bake, and a delicious salmon dinner. The sun was down on the return boat trip and the lighted Seattle skyline was both brilliantly colorful and beautiful. It was very memorable.

3.7 INSURANCE AGAINST SOLE SOURCE STOPPAGE SITUATION

Uralane 7760 was a huge improvement over the polysulfide adhesive being phased out. In fact, it was so much better that we could never go back to polysulfide again. This made us highly dependent on a supplier with proven ability to compound Uralane 7760 consistently, but less experienced in packaging it into frozen premix cartridges. It was decided that I would conduct an effort to develop a second source in-house. We now had a Ross planetary Vacuum mixer in our TSD lab, so now we should be able to do the processing properly. We pretty much knew what the successful formulation would be. It would have a butadiene polyol, an aliphatic diisocyanate, and tabular alumina powder in the formulation. There would also be a thixotropic agent, such as fumed silica, to attain non-sagging fillets. We were funded to proceed on our version.

3.7.1 DEVELOPMENT OF HACTHANE 100 ADHESIVE

The tall young man doing the lab work for this enterprise was Steve Lau. This formulating effort may have been the first sizable project that brought us together, but without his innovative participation in this and future projects, it is unlikely that I would have been as successful. He had good electrical and mechanical skills already and had both the ability and personal initiative to fabricate devices very helpful to the projects he worked on. Formulating was new to him but, he was a quick learner and soon became quite good at it.

My experience at Communication Technology Corporation, where I formulated encapsulants for telephone-buried cable splice enclosures, gave me familiarity with polybutadiene polyol/aliphatic isocyanate type polyurethanes. I had a good idea of which ingredients to use. Figure 3.1 is a likely structure for PolyBd R45HT polyol.

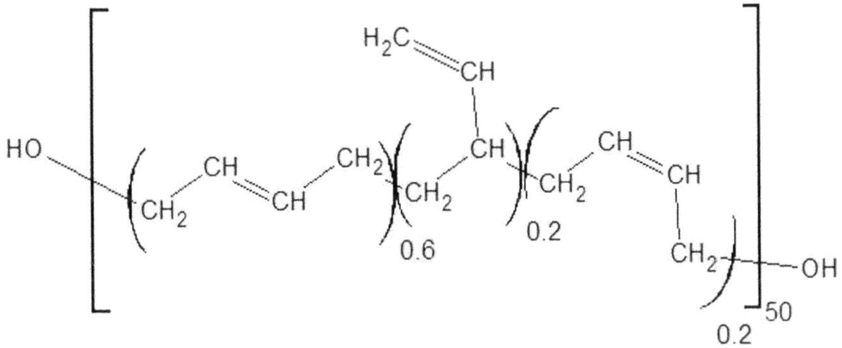

FIGURE 3.1   Likely structure for PolyBd R45HT polyol.

3.7.2 FORMULATING HACTHANE 100 FILLETING ADHESIVE

The butadiene polyol selected was PolyBd R45HT, a 3000 molecular weight triol. This ingredient would give us great electrical properties, low glass transition temperature, and the correct strength needed for a reworkable adhesive. The tabular alumina powder was already in the Hughes procurement system and could be ordered from stores. It would give us the needed thermal conductivity. The thixotrope initially used was Cab-O-Sil, also in the Hughes Stores. This ingredient gives sag resistance to a formed fillet. The aliphatic diisocyanate selected was Desmodur N3200. It would help give us long work-life and long freezer storage life. Previous experience had taught me that Dibutyl Tin dilaurate (DBTDL) was an effective catalyst for this particular combination of polyol and isocyanate. By varying the amount of DBTDL used, I could regulate the speed of the reaction. We spent several weeks battling pitted samples resulting from moisture-caused bubbles. The problem was tracked down to the Cab-O-Sil ingredient, which is highly hydrophilic and absorbs moisture out of the air. When we obtained and substituted the hydrophobic Aer-O-Sil R792 fumed silica, the problem disappeared. Note: Cab-O-Sill works fine with epoxies, where water present in Cab-O-Sil has no negative effect. Making the change from Cab-O-Sil to Aer-O-Sil R792 fumed silica was also a major breakthrough. The bubble problem disappeared and the new thixothrope worked equally well in creating a non-sag property, essential to our filleting adhesive. Without this discovery, we might have failed to meet our objective.

We had some problems consistently adding the right number of drops of DBTDL catalyst to the batches, and found that we could be more precise if we made a mixture of DBTDL in PolyBd R45HT polyol and used that mixture to weigh into the batch. Preliminary testing said we were on the right track to having our equivalent to Uralane 7760. We decided to give our filleting adhesive a name, Hacthane 100. The "HAC" part from Hughes Aircraft Company and the "thane" part of the name from urethane. In the future, there would be several more Hacthane formulations. Now, we were ready to see if Hacthane 100 frozen premix passed all of the HMS 2289 requirements.

3.7.3 THE HACTHANE 100 FORMULATION

TABLE 3.3   The Formula for Hacthane 100 Thermally Conductive Filleting Adhesive

Ingredient	Source	Weight percentage
PolyBd R45HT	Arco Chemical	24.7
325 mesh tabular alumina	HMS 20-1795	66.0
Aer-O-Sil R792 fumed silica	De Gussa	4.0
Catalyst mixture	See note 1	1.2
Desmodur N3200	Mobay chemical	4.1

Note 1: One drop of DBTDL per 5 g of PolyBd R45HT.

3.7.4 THE HACTHANE 100 COMPOUNDING PROCESS

The process for compounding and packaging Hacthane 100 adhesive is as follows:

1. Dry alumina filler overnight at 250–300°F.
2. Stir proper weight of hot alumina into PolyBd R45HT polyol.
3. Stir proper weight of Aer-O-Sil into mixture of step 2.
4. Transfer mixture from step 3 to Ross planetary mixer.
5. Close the lid and evacuate air from the Ross mixer. While stirring, raise temperature to 212–217°F using Ross mixer heating jackets. Mix for 15–30 min at 33 rpm and at 0.2 torr or less.
6. Cool mixture to room temperature using Ross mixer cooling jackets.
7. Prepare mixture of DBTDL in R45HT. Add correct amount to batch in mixer.
8. Add proper amount of Desmodur N3200 to batch in mixer.
9. Close lid, evacuate, and mix for 10–15 min at 33 rpm and 0.2 torr or less.
10. Remove mixture from Ross mixer and immediately fill 8 oz. Semco cartridges.
11. Using Semco cartridges, fill pre-labeled non-silicone small syringes.
12. Seal syringes in approved vapor barrier bags. Immediately freeze syringes in dry ice.
13. Transfer to−40°F freezer for storage.

3.8 QUALIFICATION OF HACTHANE 100 ADHESIVE

The results of testing Hacthane 100 adhesive to the requirements of HMS 2289 are listed in the following table:

TABLE 3.4 Test Results for Hacthane 100 to the Requirements of HMS 2289

Property	Test method	Requirement	Result
Work-life	Extrusion from syringe	>2 h	Pass
Shore A durometer	ASTM D2240	65–90	82
Storage life	HMS 2289	3 months minimum	9 months
Thixotropy	Vertical sag test	0.5 inch max	Pass
Specific gravity	ASTM D792	1.9–2.3	2.2

TABLE 3.4   *(Continued)*

Property	Test method	Requirement	Result
Solvent resistance See Note 2	4 cycles of isopropanol, then Freon TE vapors Per HMS 2289	13% weight gain max.	11%
Thermal conductivity, BTU/h-ft-°F	Colora	0.35 min	0.45
Glass transition temperature	Thermal mechanical analyzer (TMA)	-40°C max	−78°C
CTE > T_g	TMA	9.5 max	8.3
CTE < T_g	TMA	1.9 to 3.9	2.4
Lap shear strength, psi	ASTM D1002	100 to 350	390
Volume resistivity	ASTM D257	10^{12} ohm-cm min	1.5×10^{15} ohm-cm
Dielectric strength	ASTM D149	300 V/mil min	630 V/mil
Dielectric constant	ASTM D150	6.0 max	5.14 @ 1 KHz
			4.96 @ 10 KHz
			4.72 @ 100 KHz
			4.48 @ 1000 KHz
Dissipation factor	ASTM D150	2.6 max	1.56 @ 10 KHz
			1.62 @ 100 KHz
			1.50 @ 1000 KHz
NASA outgassing	ASTM D595	TML-WVR < 1.0%	0.06%
		CVCM < 0.1%	0.01%

Note 1: CTE > T_g means = coefficient of thermal expansion above glass transition temperature. CTE < T_g means = coefficient of thermal expansion below glass transition temperature.

Note 2: The specification was changed to eliminate Freon TE from the test as its use has been discontinued. Freon TE is deemed harmful to the ozone layer.

Hacthane 100 passed all requirements except lap shear strength, where it exceeded the 350 psi max requirement. However, we knew that it was easily reworkable, so the HMS 2289 lap shear strength limit was raised to make it acceptable. Hacthane 100 properties were similar to the properties of Uralane 7760, except its storage life was far superior and the outgassing results were better. Its glass transition temperature was −78°C versus −63°C for Uralane 7760.

A materials safety data sheet was written for Hacthane 100. This document was required before allowing it to be used for production.

The Space and Communications Group (SCG) had implemented its own specifications. Hacthane was tested and qualified to SCGMS 51014, Type 1. One thousand tubes of Hacthane 100 frozen premix were ordered by SCG in late 1992.

3.9   THE ABLESTIK LABORATORIES CANDIDATE TO HMS 2289

SCG's manufacturing department took the initiative to request that Ablestik Laboratories develop and submit a filleting adhesive, which meets the requirements of HMS 2289. They submitted a sample having the designation XP-HMS-2289 for our evaluation. We decided to test it for outgassing and then decide whether to continue qualification testing. It passed with a TML-WVR = 0.40 and a CVCM = 0.01.

However, we noticed several problems when attempting to cure the material:

(1) Bubbling in the material indicated a moisture contamination problem.
(2) The material failed to properly cure within 1 h at 200°F. An additional hour made little difference.
(3) The material clearly failed the sag resistance requirement, which is essential for a filleting adhesive.

Ablestik was told of the problems and asked to try again. The second sample had the same deficiencies.

3.10   PATENT FOR GENERIC POLYURETHANE FILLETING ADHESIVES AND OTHER APPLICATIONS

Steve Lau and I filed for the patent on January 28, 1994. HAC encouraged us to earn patent awards. Monetary awards totaling over $1000 could be earned if one's patent application made it through the process and was finally granted. Moreover, a team of patent attorneys housed in the Hughes Corporate Offices were there to help us and others throughout the corporation. TSD was a large contributor to HAC's largess of patents.

Consequently, one of the patent attorney technologists, Mary Lachman, actually worked with TSD on a full-time basis. She was essential to our patent success over the years. She was dedicated to help us run the gauntlet of the U.S. Patent Office, where it seems their goal is to disqualify the submitted patent applications. There is a lot of work involved in preparing the patent application and defending it. Several years of back and forth legal battling typically goes on. TSD management was stingy with their charge numbers. We ended up donating many tens of hours of work answering the questions Mary Lachman had as she translated our explanations into patent legalese.

The title of our patent was "Frozen premix, fillet-holding urethane adhesives/sealants." The patent application was filed on January 28, 1994 and the patent was granted very close to 1 year later on January 31, 1995. This one was surprisingly fast; many of the others took years. The U.S. Patent number is 5,385,966. This patent has expired and anyone can use what it teaches. I will discuss possible spin-off ideas in the latter part of the book.

KEYWORDS

- **space electronics**
- **polysulfide adhesive**
- **thermal conductivity**
- **NASA**
- **Hughes materials specification**

CHAPTER 4

FLEXIBLE EPOXY THERMAL TRANSFER ADHESIVES FOR FLATPACKS

CONTENTS

In the previous chapter, the major part of the polysulfide replacement effort was accomplished with the introduction of Uralane 7760 and Hacthane 100 filleting adhesives. In this chapter, we continue the effort by finding a suitable replacement for the polysulfide film adhesive used under flat-bodied components. The purpose of the thermally conductive film adhesive is to provide a thermal path for the flat-bodied components to dissipate heat and run cooler. The electronics in satellites operate in a vacuum, so there is no air available to blow over them as is normal for earth-bound electronics. The goal is to maximize the life of these components by allowing them to run cooler.

4.1   BACKGROUND TO THE FLATPACK ADHESIVE APPLICATION

Remember that the filleting adhesive was 85% of the polysulfide adhesive usage. The other 15% usage was as a thermal transfer adhesive in either (a) frozen premix film form or (b) frozen premix tear strip form. The film form consisted of a glass cloth carrier coated on each side with thermally conductive adhesive. The tear strip form was a monofilament Kevlar fiber arranged in a circuitous path and coated with thermally conductive adhesive. The idea was that in order to remove a bonded flatpack from a printed wiring board, the worker located the end of the fiber and pulled on it to tear the bondline apart. The film and tear strip adhesives were processed in such a manner that the thickness was correct for fitting under flat-bodied components, and thus did not induce stress in the flatpack leads. Release film was applied to both sides of the adhesive during its fabrication. When the assembly technician was ready to apply the film or tear strip adhesive, he/she cooled the adhesive on a block of dry ice, which allowed the release film to easily come off.

4.1.1   FILM ADHESIVE USING URALANE 7760 PASTE

Alumina-filled polysulfide adhesive worked quite well in the fabrication and application of these film or tear strip forms, except of course for the fact that it failed the outgassing test badly. Now, we hoped that Uralane 7760 would work as well. In order to confirm our hopes, we used Uralane 7760 adhesive to make the film adhesive version. Unfortunately, we ran into a problem that we were unable to resolve due to the basic chemistry of polyurethane compounds. When we removed the release film using dry ice, moisture from the air wicked into the exposed paste adhesive surface. When the film adhesive was cured at 200°F, bubbles formed in the cured adhesive. This is due to reaction of water with the isocyanate groups in the adhesive. The product of this reaction is carbon dioxide and a primary amine. The primary amine immediately reacts with another isocyanate group to form a urea linkage. The carbon dioxide gas expanded during the heat cure and produced bubbles in the cured adhesive. Of course, bubbles are the last thing we want to have in an adhesive, which was intended to conduct heat away from the components. After many failed attempts to make Uralane 7760 film adhesive work, we realized that there was no remedy to the problem. We needed to find a different way to replace polysulfide in the film adhesive form.

4.2   THE FLEXIBLE EPOXY APPROACH TO A FORMULATION

Sometimes you have to think outside the box. Polysulfide polymers did not work due to outgassing problems. Polyurethanes did not work due to bubbling. What other polymers should we consider? I knew that epoxy resin formulations could be rubbery. I personally handled a very flexible cured epoxy sample back in my days at Argonne National Laboratories (ANL) 25 years previously. I further mused, "Perhaps, a flexible epoxy formulation could work for the film adhesive." I felt that Steve Lau and I needed to explore that possibility. It seemed to me to be the most likely way to find a replacement for polysulfide in the film adhesive form. I made a list of my arguments for attempting to formulate a flexible epoxy film adhesive:

(1) Epoxies do not have a foaming problem from moisture as do polyurethanes.
(2) Epoxies are excellent adhesives to a wide variety of different materials.
(3) Flexible epoxies can be made cohesively weak to facilitate removal of components.
(4) Epoxies have long storage lives as frozen premix compounds.
(5) Epoxies are dielectric materials.

4.2.1   EMPLOYING FLEXIBLE EPOXY RESINS

I knew that a few flexible epoxy resins were commercially available to the formulator. However, the conventional wisdom among epoxy formulators is that these flexible resins could be used at no greater than 10–20% of an epoxy resin blend, where rigid epoxy resins constitute the major part of that blend. The advantage gained from this blend is that the resulting plastic will be less brittle than it would be without the flexible resin. On the other hand, the conventional wisdom advised against higher loadings, because the properties of the resultant epoxy plastic would be useless. The idea of using the flexible resins as 100% of the resin blend was unthinkable. Of course, that is exactly what I wanted to do.

TABLE 4.1   Flexible Epoxy Resins Used in this Effort

Supplier	Epoxy resin trade name	Epoxide equivalent	Chemical structure
Dow Chemical	DER 732	310-330	Diglycidyl ether of poly(oxypropylene) glycol
Dow Chemical	DER 736	175-206	Diglycidyl ether of dipropylene glycol
Shell Chemical	Epon 871	430	Diglycidyl ether of Linoleic dimer acid
Shell Chemical	Epon 872	700	Adduct of 2 moles DGEBA and 1 mole of linoleic dimer acid (see note 1)

Note 1: DGBDA is common abbreviation for the diglycidyl ether of bisphenol A. Shell Chemical's Epon 828 is an example of a DGBDA epoxy resin.

Note 2: Some of the chemical structures are shown below:

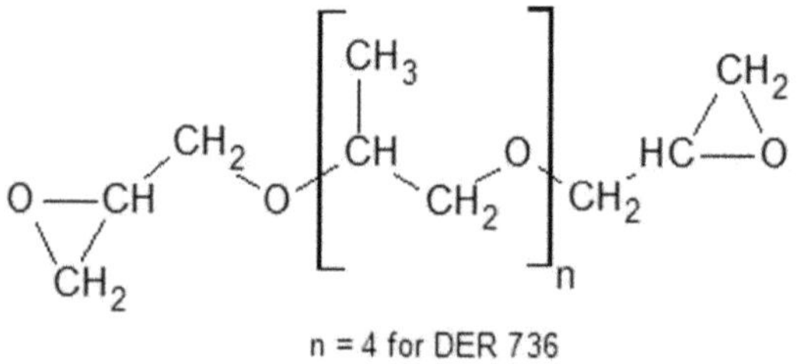

n = 4 for DER 736
n = 9 for DER 732

FIGURE 4.1   Chemical structure for DER 732 and DER 736 epoxy resins.

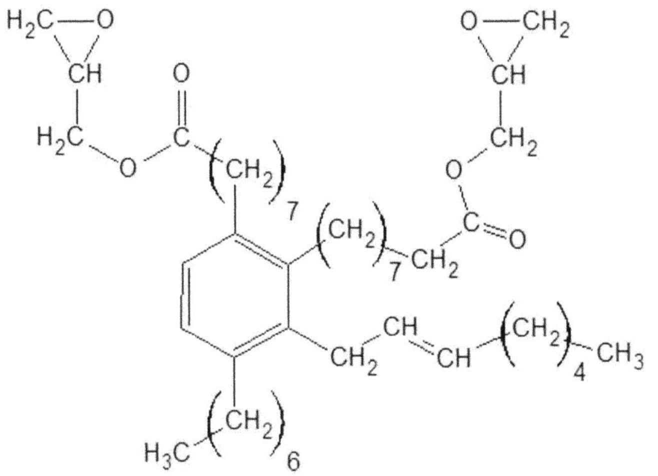

FIGURE 4.2   Chemical structure for Epon 871 epoxy resin.

4.2.2   *EMPLOYING FLEXIBLE CURATIVES FOR THE FLEXIBLE EPOXY RESINS*

There are a wide variety of curing agents available to cure epoxy resins. Many of them such as aromatic diamines or acid anhydrides can be ruled out because they would fail to cure the resultant adhesive in the required 20 min at 200°F. However, there is one particular category of curatives, which could meet this requirement, namely, aliphatic diamines or polyamines.

Some of these aliphatic amines would also impart flexibility to the cured epoxy compound. I wanted to make the resin/curing agent system result in as flexible a material as I could manage. I agree that the mechanical properties would likely be weak; however, for this application, weakness is a benefit. Weakness in the adhesive is beneficial to make component removal all the easier. The following table lists the curatives, which were evaluated:

TABLE 4.2 Fast-Setting Curatives Used in this Effort

Supplier	Curative trade name	Equivalent Wt. per active H	Chemical structure
Air products	Ancamine T-1	47	N(2-hydroxy ethyl) diethylene triamine
Several	Hexane diamine	29	1,6-Hexane diamine
DuPont	Dytek A	29	2-Methyl-pentamethylene diamine
Dow Chemical	AEP	43	Amino-ethyl-piperazine

Note: Ancamine AD and Ancamine XT were also evaluated and found to be too fast curing.

4.2.3 ESTABLISHING FEASIBILITY OF THE CONCEPT

David Vachon and Steve Lau evaluated several mixtures of flexible epoxy resin and aliphatic polyamine curatives beginning in December 1987. Some formulations were alumina filled and others were unfilled. By late February 1988, they had evaluated 12 different formulations and most of them were too hard and strong, but a few of them were rubbery. The concept was proven feasible. The remaining thirty-some formulations to be evaluated were aimed at meeting the application requirements. Almost all of those trials were rubbery and cured within 20 min at 200°F.

Our flexible epoxies would allow us to replace polysulfide as a thermally conductive film adhesive under flatpacks. We were further convinced that these rubbery epoxies potentially had widespread application in aerospace electronics. However, we faced a communication problem. If we told design engineers that these adhesives were epoxies, they would associate the word "epoxy" with the rigid, brittle epoxies from their experience. To prevent that stereotyping, we came up with the trade name "Flexipoxy." This simple change in terminology worked and the word "Flexipoxy" became a common term at HAC.

4.3 STEERING COMMITTEE FOR THE REPLACEMENT OF POLYSULFIDE FILM ADHESIVE

A steering committee continued from the polysulfide replacement effort, which resolved the filleting adhesive replacement effort. It consisted of three departments: (1) SCG Digital Electronics Engineering, (2)

SCG Manufacturing, and (3) Technology Support Materials and Process Engineering. Now that a feasible path to success in replacing the polysulfide film adhesive existed, the steering committee became active again. The kick-off meeting for the film adhesive effort was held on April 1, 1987. The chairman of the committee was R.B. Mitsuhashi. There were three other members addressing design concerns, another three members addressing manufacturing concerns, and three members from TSD. I was one of them from TSD and had the responsibility of optimizing the new adhesive so that it could be qualified and put into production.

4.3.1 INITIAL FORMULATING EFFORT

The screening of trial formulations was mostly qualitative rather than quantitative because we didn't know what the results would be like with these novel materials. The screening criteria used are listed in the following table:

TABLE 4.3   Screening Criteria for Flexible Epoxy Candidates

Property	Test method	Requirement
Cure time at 200°F	Observation and probing	10–20 min
Flexibility	Deformation of cured specimen	Rubbery
Tensile strength	Deformation of cured specimen	Not too strong to rework and not too weak for survival
Volume resistivity	ASTM D257	$>1 \times 10^{12}$ ohm-cm*
NASA outgassing	ASTM D595	TML-WVR to be 1.0% max.
		CVCM to be 0.1% max.

*Requirement was later lowered to 1×10^{11} ohm-cm.

There was a requirement that the final compound has a work-life of at least 2 h, yet cure within 20 min at 200°F. Each trial formulation was tested for ability to cure within 20 min. We were hoping for a very rubbery epoxy once cured. Stretching and releasing the cured specimens gave us an idea of how soft and lively the trial compounds were. Both filled and unfilled formulations were evaluated and the feel is quite different for each type. The very high filler level in a thermally conductive compound tends to stiffen the compound, whereas the unfilled compound may be very flexible. Next, we tried to break the cured specimens by stretching them. We were looking for an intermediate strength between too friable or weak and too strong to allow component removal without damage to the component.

SELECTIVE TESTING

Candidates that looked promising were tested for volume resistivity. Values of 1×10^{12} ohm-cm were of further interest. Finally, very promising candidates were tested for NASA outgassing.

ELIMINATION OF RESINS OR CURATIVES

Forty-six different flexible epoxy formulations were examined before we had a formulation worthy of submitting to the qualification testing. Flexible epoxy resins were identified and selected for evaluation. All of the flexible epoxy resins (DER 732, DER 736, Epon 871, and Epon 872) used as 100% of the epoxy resin or blended with each other cured to a rubbery consistency with the curatives listed. It was necessary to use blends of flexible epoxy resins in order to maximize dielectric properties.

All but two of the curatives were eliminated based on having acceptable cure times at a cure temperature of 200°F. The finalist curatives were Ancamine T-1 and hexane diamine. Eliminated were Dytek A and aminoethyl-piperazine, which needed 30 min or more to cure the compound.

DE-AIRING FILLED COMPOUNDS

Until we finally acquired a Ross planetary mixer, which allowed heating, cooling, and mixing of viscous materials under vacuum, we needed a process to remove mixed in air from the filled compounds. It was found that centrifuging the compounds was an effective method.

4.3.2   FINAL STAGE OF THE FORMULATING EFFORT

FORMULATING FOR EASY COMPONENT REMOVAL

A combination of 10 parts DER 871, 18 parts DER 736, 3.5 parts hexane diamine, 63.5 parts alumina powder, and 2.5 parts Cab-o-Sil fumed silica produced a thermal transfer adhesive with promising properties. However, it was deemed too strong for safe component removal.

I knew from my formulating experience at chemical Products Corporation (CPC)that unreactive plasticizers reduced cohesive strength markedly. However, I needed to select one that would remain tied up in the polymer matrix so it did not hurt outgassing properties. I concluded that a very long chain polyoxypropylene triol, like Union Carbide LHT-34, might do the job. As a liquid, it would be miscible with the DER 736 epoxy resin due to their common polyoxypropylene backbone. The LHT-34 would be retained in the cured polymer for the same reason. In retrospect, I believe that some of the LHT-34 becomes part of the growing polymer due to linking via its hydroxyls groups.

The following two tables show the formulations and properties of the finalist candidates:

TABLE 4.4   Formulations of the Finalist Candidates

Ingredient	Trial 26	Trial 38	Trial 41
DER 736 epoxy resin	15.6 pbw	20.0 pbw	18.0 pbw
Epon 871 epoxy resin	9.0 pbw	11.6 pbw	10.0 pbw
LHT-34 triol	2.4 pbw	3.1 pbw	2.5 pbw
Glass beads	9.0 pbw	None	none
Hexane diamine	3.0 pbw	3.9 pbw	3.5 pbw
Alumina powder	59.5 pbw	60 pbw	63.5 pbw
Cab-O-Sil powder	1.5 pbw	1.5 pbw	2.5 pbw
Totals	100.0 pbw	100.0 pbw	100.0 pbw

TABLE 4.5   Properties of the Finalist Candidates

Properties	Trial 26	Trial 38	Trial 41
Cure time at 200°F	15-20	15-20	15-20
Flexible	Yes	Yes	Yes
Tensile strength	Low	Low	Low
Volume resistivity, ohm-cm	5.9×10^{12}	Not tested	4.5×10^{12}
Outgassing			
TML-WVR	0.66	0.65	0.62
CVCM	0.04	0.05	0.06

Note: The glass beads in Trial 26 were deemed unnecessary and are no longer in Trials 38 and 41. The improvements between Trial 38 and Trial 41 have to do with ease of making a film adhesive. Trial 41 was advanced to full qualification testing and was named Flexipoxy 100 adhesive.

4.3.3   MAKING A FUNCTIONAL FILM ADHESIVE

STEVE LAU'S INNOVATIONS

Steve Lau did all of the process development effort necessary to make a suitable film adhesive using Flexipoxy 100 adhesive. One innovative idea was to impregnate the 2 mil thick glass cloth carrier with Flexipoxy 100 paste and cure it. Essentially, the impregnated glass cloth was covered on each side with release film and placed in a platen press. The temperature of the platens was raised to 200°F to cure the adhesive.

The advantages of this idea were: (1) it made the cloth easier to handle during the subsequent coating processes, and (2) the cured cloth became a physical barrier to metallic contaminants or solder spikes, which might cause shorts.

BUILDING OUR OWN FILM-PULLING MACHINE

Once Steve Lau understood the concept of a film-pulling machine, he ordered the parts and materials and fabricated a machine of his own design. TSD had a machine shop available within our laboratory for such prototype work. The film-pulling machine used a chain and sprocket arrangement to pull the film adhesive under a doctor blade.

4.4 THE FINALIST—FLEXIPOXY 100 FILM ADHESIVE

4.4.1 *THE FLEXIPOXY 100 PASTE ADHESIVE FORMULATION*

Table 4.6 below lists the formulation for Flexipoxy 100 paste adhesive, which is used to make the film adhesive.

TABLE 4.6   The Formulation for Flexipoxy 100 Paste Adhesive

Ingredient	Supplier	Description	Weight percent
Epon 871	Shell Chemical	Flexible epoxy resin (fatty acid based)	10.0
DER 736	Dow Chemical	Flexible epoxy resin (poly-oxypropylene based)	18.0
Hexane diamine	Various	Primary diamine	3.5
LHT-34 triol	Union Carbide	Long chain polyoxypropyl-ene triol	2.5
Cab-O-Sil M5	Cabot	Fumed silica powder	2.5
T-61 alumina powder	Alcoa	325 mesh aluminum oxide powder	63.5
Total			100.0

4.4.2 *DESCRIPTION OF FLEXIPOXY 100 FILM ADHESIVE*

The thickness of the film adhesive is controlled to be between 12 and 18 mils. The film adhesive consists of a central layer of Flexipoxy 100—impregnated, 2 mil thick, glass cloth. The impregnated glass cloth was cured to facilitate later processing. The central layer is then covered on each face with a layer of Flexipoxy 100 paste adhesive. The adhesive layers are thickness controlled by using a film puller and doctor blade apparatus. Outside of each paste adhesive layer is a thin plastic film, which is removed prior to use of the film adhesive. They are called release films.

The completed film adhesive is then immediately frozen to stop the polymerization reaction. The film adhesive is kept in a frozen premix state until ready to apply to hardware for bonding to the circuit board.

The process for making the film adhesive is high tech indeed. Great pains are taken to ensure that no conductive fibers, such as graphite, exist in the film adhesive. Such contaminants could be damaging to the electronics. This includes using fume hoods and laminar flow clean rooms during preparation of the adhesive.

4.5   QUALIFICATION TESTING OF FLEXIPOXY 100 ADHESIVE

4.5.1   OUTGASSING PROPERTIES OF FLEXIPOXY 100 PASTE ADHESIVE

Table 4.7 below shows the outgassing properties of Flexipoxy 100 in both paste and film adhesive form. Polysulfide paste results are shown for comparison.

TABLE 4.7   Outgassing Properties of Flexipoxy 100 Paste and Film Adhesive

Property	Method	Requirement	Flexipoxy 100 paste adhesive	Flexipoxy 100 film adhesive	Polysulfide adhesive
Outgassing TML-WVR:	ASTM D595	Less than 1.0%	0.59	0.58	23.1
CVCM:		Less than 0.1%	0.08	0.08	0.49

Comment:

This table is the most important one of them all! The polysulfide adhesive does not just fail outgassing, it is literally decomposing. Consider that about 70 weight percent of the adhesives is alumina and silica, which are thermally stable. Therefore, about two-thirds of the polysulfide polymer is lost during the outgassing heat and vacuum exposure. By contrast, Flexipoxy 100 adhesive is thermally stable and passes the outgassing requirements.

4.5.2   PROCESSING PROPERTIES OF FLEXIPOXY 100 FILM ADHESIVE

The processing properties of Flexipoxy 100 paste and film adhesives are shown in the following table:

TABLE 4.8 Processing Properties of Flexipoxy 100 Paste Adhesive

Property	Method	Requirement	Flexipoxy 100 Paste Adhesive	Flexipoxy 100 Film Adhesive
Color	Visual	Off-white	Off-white	Off white
Specific gravity	Weight volume	2.0–2.2	2.10	Not tested
Film thickness	Micrometer	12–18 mils	Not applicable	Pass
Ease of release film removal	Manufacturing evaluation	Easy	Not applicable	Pass
Work-life	Extrusion from syringe	>2 h	4 h	Not applicable
Work-life	Torsional shear	None established	4 h	4 h
Thixotropy	Visual	Negligible sag or migration	Pass	Pass
Compatibility with Uralane 5750	Application test	None	Pass	Pass
Compatibility with Uralane 7760	Application test	None	Pass	Pass
Freezer storage life at −40°F	(1) Extrusion test	(1) >2 h	4 h	Not applicable
	(2) Torsional shear	(2) >2 h	4 h	4 h
Ease of component removal	Manufacturing evaluation	None established	Pass	Pass
Cure behavior at 200°F	Shore A durometer	For info only:		
	Per ASTM D2240	At 10 min	66	
		At 20 min	79	
		At 30 min	82	
		At 1 h	81	
		At 2 h	82	

Comments:

Uralane 5750 is a conformal coating which is applied to the cured thermal transfer adhesive. Uralane 7760 is the new filleting adhesive. This compatibility test did not involve curing both Flexipoxy 100 and Uralane 7760 together in contact with each other. A curing problem was discovered months later (see Part II Chapter 3).

4.5.3   MECHANICAL PROPERTIES OF FLEXIPOXY 100 PASTE ADHESIVE

Table 4.9 shows the durometer of Flexipoxy paste adhesive and lap shear strength of various substrates when bonded by Flexipoxy 100 film adhesive.

TABLE 4.9   Mechanical Properties of Flexipoxy 100 Paste Adhesive

Property	Method	Requirement	Flexipoxy 100 paste adhesive	Flexipoxy 100 film adhesive
Durometer,	ASTM D2240			
Shore A		80–90	82	Not applicable
Lap shear strength Aluminum-to-aluminum	ASTM D1002			
At 75°F:		100–500 psi	480 psi	385 psi
At 200°F:		Info only	360 psi	256 psi
At 60°F:		Info only	2160 psi	1980 psi
Lap shear strength	ASTM D1002			
(1) Nickel plate (Ajax scrubbed)		Info only		430 psi
(2) Nickel plate (Freon TE cleaned)				410 psi
(3) Solder plate (Freon TE cleaned)				510 psi
(4) Polyimide (Freon TE cleaned)				430 psi

Comments:

8 mil film adhesive had a lap shear strength of 385 psi at room temperature, whereas 15 mil film adhesive had a lap shear strength of 510 psi on aluminum-to-aluminum specimens.

4.5.4 THERMAL PROPERTIES OF FLEXIPOXY 100 PASTE ADHESIVE

Table 4.10 shows the thermal conductivity, glass transition temperature, dynamic shear modulus, and coefficients of thermal expansion for Flexipoxy 100 paste and film adhesives.

TABLE 4.10   Thermal Properties of Flexipoxy 100 Paste Adhesive

Property	Method	Requirement	Flexipoxy 100 paste adhesive	Flexipoxy 100 film adhesive
Thermal conductivity BTU/h-ft-°F	ASTM C177	>0.25	0.332	0.300
Glass transition temperature	Thermomechanical analysis	+4°C	−4.1°C	−4.1°C
Glass transition temperature	Cured-sample rheometrics	+4°C	−6°C	−12°C
Dynamic shear modulus	Cured-sample rheometrics	Info only		
At 50°C:			2800 psi	4600 psi
At 100°C:			2800 psi	4500 psi
CTE above Tg, ×10⁻⁵ in/in/°F	Thermomechanical analysis	Less than 9.5	7.5	7.5
CTE below Tg, ×10⁻⁵ in/in/°F	Thermomechanical analysis	1.9 to 3.9	2.4	2.4

Comments:

It would have been better if our flexible epoxy had had a lower glass transition temperature. In Part II Chapter 4, we would formulate for that improvement.

4.5.5 ELECTRICAL PROPERTIES OF FLEXIPOXY 100 PASTE ADHESIVE

Table 4.11 shows the volume resistivity, dielectric strength, dielectric constant, and dissipation factor for Flexipoxy 100 paste and film adhesives.

TABLE 4.11   Electrical Properties of Flexipoxy 100 Paste Adhesive

Property	Method	Requirement	Flexipoxy 100 paste adhesive	Flexipoxy 100 paste adhesive
Volume resistivity, ohm-cm	ASTM D257 At 500 V	$>1.0 \times 10^{11}$	5.6×10^{11}	5.7×10^{11}
Dielectric strength, V/mil	ASTM D149	Info only	658	658
Dielectric constant	ASTM D150	Info only		
At 1 KHz:			10.7	11.3
At 10 KHz:			9.7	9.8
At 100 KHz:			8.3	8.4
At 1000 KHz:			7.1	7.1
Dissipation Factor	ASTM D150	Info only		
At 10 KHz:			8.1%	9.6%
At 100 KHz:			9.5%	10.6%
At 1000 KHz:			8.3%	10.6%

Comments:

It would have been better if our flexible epoxy had better dielectric properties. In Part II Chapter 4, we would make big improvements in this regard.

4.5.6   CURING BEHAVIOR OF FLEXIPOXY 100 ADHESIVE

The curing characteristics of Flexipoxy 100 adhesive were studied using the rheometrics analyzer. Using this device, it was possible to determine the torsional shear strength of the curing adhesive versus time at different cure temperatures. The results are presented in the following table:

TABLE 4.12   Torsional Shear Strength Increase versus Cure Time for Flexipoxy 100 Adhesive

Cure temperature (°F)	15–25 in-lb	35–45 in-lb	55–65 in-lb	>85 in-lb
75	16 h	48 h	72 h	>96 h
100	3.5 h	6 h	10 h	16 h
120	80 min	120 min	140 min	180 min
140	30 min	45 min	60 min	90 min
160	15 min	30 min	45 min	75 min
180	10 min	18 min	30 min	45 min
200	5 min	10 min	20 min	25 min

Flexipoxy 100 adhesive has the ideal combination of adequate work-life at room temperature and fast cures at oven temperatures at or below 200°F. A torsional shear strength of 85 in-lb or greater was considered to be a full cure. A work-life of 2 h minimum is required of the adhesive. Flexipoxy 100 had a 4-h work-life.

4.5.7 STORAGE LIFE TESTING OF FLEXIPOXY 100 PASTE AND FILM ADHESIVES

An extensive study was also run to establish the stability of Flexipoxy 100 paste adhesive and Flexipoxy 100 film adhesive during freezer storage. The property requirements of HMS 2308 and HMS 2306 were determined initially and after 3 months at a freezer storage temperature of −40°C. The test results confirmed that both paste and film adhesive are stable for at least 3 months at freezer conditions.

4.5.8 BATCH-TO-BATCH CONSISTENCY OF FLEXIPOXY 100 PASTE AND FILM ADHESIVES

Four different batches of Flexipoxy 100 film adhesive were prepared and run through an extensive series of tests to establish that we could manufacture the adhesive consistently. Tests included film thickness, outgassing, solvent resistance, glass transition temperature, coefficient of thermal expansion (9 above and below Tg), dielectric strength, dielectric constant, dissipation factor, and lap shear strength,. In addition, manufacturing compared thermal cycling and removability, Uralane 5750 coating compatibility, adhesive wetting, work-life, storage life, and ease of processing for the four batches.

All testing confirmed that the four batches met the requirements and were very consistent.

4.5.9 NEW SPECIFICATIONS

Three new specifications were prepared: HMS 2308 defines Flexipoxy 100 paste adhesive or any future candidate with equivalent properties. HMS 2306 defines Flexipoxy 100 film adhesive or any future candidate with equivalent properties. Finally, HPR 63031 defines the application and processing for the film adhesive.

4.5.10   DISCONTINUED INGREDIENT PROBLEM

LHT-34 TRIOL REPLACED BY LHT-28 TRIOL

Union Carbide, the supplier of the ingredient LHT-34, withdrew the product from the market due to low usage. Fortunately, LHT-28 triol was available to replace it. The following table summarizes the results of our evaluation:

TABLE 4.13   Comparison of Adhesive Properties Using LHT-34 versus LHT-28 Triol Ingredient

Property	Adhesive form	Adhesive using LHT-34	Adhesive using LHT-28	Comments
Plasticizer compatibility	Paste	Good	Good	Acceptable
Work-life, using lap shear strength on 2-h-old adhesive	Paste	410 psi SD = 40 psi	410 psi SD = 40 psi	Acceptable
Volume resistivity, x 10^{11} ohm-cm	Film	2.9–6.8	3.4–4.6	Acceptable
Dielectric strength, V/mil	Paste	702	688	Acceptable
	film	545	527	Acceptable
Outgassing: TML-WVR:	Paste	0.76%	0.73%	Acceptable
CVCM:		0.04%	0.02%	
Solvent resistance: Weight:	Paste	+ 4.43%	+ 4.53%	Acceptable
Volume:		4.33%	+ 4.41%	
Lap shear strength	Film	250 psi SD = 50 psi	300 psi SD = 30 psi	Acceptable

The replacement of LHT-34 with LHT-28 was deemed acceptable and the specifications were revised to authorize the change.

4.6   ABLESTIK ALLOWED TO MAKE FLEXIPOXY 100 UNDER AGREEMENT

Manufacturing wanted a second source for the film adhesive other than TSD. Ablestik had already been supplying frozen premix compounds to them for many years and was a desirable source. However, they were unlikely to be able to formulate their own proprietary compound with all of the properties that Flexipoxy 100 had, so I suggested to the steering committee that the fastest way to qualify Ablestik to HMS 2306 and HMS 2308 was to provide them with the formulations and processes for making the paste and film adhesive. The members agreed and an agreement was negotiated with Ablestik.

By November 1988, we had received sample batches from Ablestik and conducted qualification testing. The test results matched those of Flexipoxy 100 and more importantly, passed the specification requirements. Ablebond 8220K paste adhesive was qualified to HMS 2308 and AbleFilm 5010K was qualified to HMS 2306.

4.7   SAMPE PAPER AND PRESENTATION

Four of us co-authored a paper describing this successful project. The authors were Robert B. Mitsuhashi, James C. Cammarata, Matthew T. Mika, and myself. Bob Mitsuhashi was the manager of the Product Development Department of the Space Vehicles Electronics Division in SCG. He was also chairman of the steering committee for this project. Matt Mika was a Senior Staff Engineer, and Jim Cammarata was an MTS in Bob's department. They were also members of the working committee representing Digital Electronics Engineering. The title of our paper was, "Development of a Thermal Transfer Adhesive For Space Electronics." I presented the paper on June 13, 1990 at the SAMPE Electronics Materials and Processes Conference in Albuquerque. The balloon festival was underway with a hundred or more colorful balloons aloft over the city. It was a delightful trip for me. I ran into many of my old colleagues still working at Sandia National Labs. Some of them were also presenting papers at the conference.

4.8   PATENTS

David Vachon, a PhD chemist working in our sister department, had time available to assist on the formulation development effort and prepared batches of experimental flexible epoxy according to my instructions. He also volunteered to prepare the patent and I was happy to delegate that task and focus on more pressing matters. Somehow, his name was listed first during the patent application drafting process.

Patent U.S. 4,866,108, titled "Flexible epoxy adhesive blend" was awarded to D. Vachon, RD Hermansen, and S.L. Lau on September 12, 1989.

KEYWORDS

- flatpack
- Uralane 7760
- epoxy resins
- Flexipoxy 100

CHAPTER 5

MORE ABOUT THE NEW THERMAL TRANSFER ADHESIVES

CONTENTS

This is the only chapter in Part II of the book which does not result in a new patent. Instead, this chapter is about what else happened with the two polysulfide adhesive replacements after they were introduced. The new adhesives were now being used in place of the polysulfide filleting and film adhesives they replaced. Uralane 7760 was the first replacement introduced, but after the near-equivalent Hacthane 100 adhesive was okayed for production, it was preferred. Flexipoxy 100 film adhesive replaced the polysulfide film adhesive. Both TSD and Ablestik could now supply it. However, the story did not end there. Engineering still wanted to be sure that they had not overlooked anything and a problem popped up in manufacturing that needed investigation and resolution.

I have included the stories of these follow-on projects in the book so that readers can see an example of how aerospace projects can become exacting and ultraconservative. Remember that a satellite can be worth hundreds of millions of dollars and the cost to put it in orbit is very expensive too. It is unacceptable to learn of your mistakes after the satellite is in orbit. Then it is too late to do anything about them.

This chapter contains the descriptions of three separate tasks conducted on the new polysulfide replacement adhesives.

Task 1 has us developing additional material property data for a finite element analysis. Task 2 has us conducting a simulated space aging study to predict the life expectancy of the adhesives, and Task 3 has us attempting to understand what happens when Flexipoxy 100 and Uralane 7760 are mixed together in the uncured state.

5.1   TASK 1: DATA FOR FINITE ELEMENT ANALYSIS

5.1.1   SPECIAL TESTING OF FLEXIPOXY 100 AND URALANE 7760

During the qualification process for Flexipoxy 100 film adhesive, it was deemed advisable to have stress analysts run a finite analysis of flatpacks bonded using the new adhesives. In addition to the use of Flexipoxy 100 film adhesive, some programs had used Uralane 7760 paste adhesive in conjunction with G-10 laminate to provide a physical barrier. The temperature range of interest is −34–99°C. Dwight Swett was assigned the stress analysis problem and I was assigned to develop whatever property data he needed.

What property data were requested are shown in the following sections. Three adhesives were tested, namely, Flexipoxy 100 film adhesive, Flexipoxy 100 paste adhesive, and Uralane 7760 paste adhesive. Comparative data for these adhesives are shown in the following tables:

5.1.2 TENSILE AND COMPRESSIVE PROPERTIES

The most basic properties needed for finite element analysis are the tensile and compressive properties as well as Poisson's ratio. Table 5.1 summarizes these properties for the three adhesives.

TABLE 5.1   Tensile and Compressive Properties plus Poisson's Ratio for Flatpack Adhesives

Property	Flexipoxy 100 film adhesive	Flexipoxy 100 paste adhesive	Uralane 7760 paste adhesive
Tensile strength, psi, per ASTM D 412	3284	553	458
	SD = 222	SD = 335	SD = 30
	Avg. of 6 specimens	Avg. of 8 specimens	Avg. of 6 specimens
Tensile modulus, psi, per ASTM D 412	542,000	Undefined	194
	SD = 140,000		SD = 59
Ultimate elongation, %, per ASTM D 412	4.19	11.7	286
	SD = 1.46	SD = 2.4	SD = 34
Compressive strength, psi, per ASTM D695	1810	1878	1122
	SD = ?	SD = 40	SD = 39
	Avg. of 1 specimen	Avg. of 2 specimens	Avg. of 2 specimens
Compressive modulus, psi, per ASTM D695	40,000	15,000	10,000
	SD = ?	SD =5000	SD = 0
Poisson's ratio	0.219	0.405	0.398
	SD = ?	SD = 0.027	SD = 0.008

Comments:

1. Flexipoxy film adhesive: Adhesive was pulled in the plane of the glass cloth carrier for tensile property determination at a crosshead speed of 0.05 in/min. The paste adhesives were pulled at 20 in/min.

2. Flexipoxy paste adhesive: Secant tensile modulus is undefined for materials having less than 100% ultimate elongation.

3. Uralane 7760 adhesive: Secant modulus is modulus at 100% elongation.

4. The compression specimens were used to determine Poisson's ratio. Poisson's ratio for an ideal elastomer is 0.5. Such elastomers deform while maintaining constant volume. The fact that these thermal transfer adhesives are highly filled with alumina particles may contribute to their deviance from a Poisson's ratio of 0.5. Yet, the two paste adhesive's values are close to 0.5. Flexipoxy 100 paste is 0.405 and Uralane 7760 is 0.398. Although they are different polymer based, their Poisson's ratios are nearly identical. By contrast, the film adhesive is anisotropic and constrained by the woven glass cloth within it.

5.1.3 DOUBLE LAP SHEAR STRENGTH

The shear strength of Flexipoxy 100 film adhesive and Uralane 7760 at −34°C was important for the stress analysis. The table shows the results:

TABLE 5.2   Double Lap Shear Strength at −34°C

Adhesive	Double lap shear strength, psi
Flexipoxy 100 film adhesive	>2100
Uralane 7760 paste adhesive	320
	SD = 60

Comments:

1. Seven specimens for each adhesive were tested.

2. The Flexipoxy 100 film adhesive at −34°C is a reinforced plastic, and was therefore stronger than the PWB, which failed at 2100 psi in tension. The film adhesive is much stronger than 2100 psi.

3. The relatively lower strength of Uralane 7760 is due to it still being a soft rubber at a low temperature of −34°C.

5.4   GLASS TRANSITION TEMPERATURE AND THERMAL EXPANSION COEFFICIENTS

The glass transition temperature, Tg, and coefficients of thermal expansion, CTE > Tg and CTE < Tg, of the three adhesives were determined using a Harrop Model 720 quartz dilatometer.

TABLE 5.3   Thermal Expansion Properties of Adhesives by Quartz Dilatometry

Adhesive	Run	Glass transition temperature, °C	CTE above Tg, ppm/°C	CTE below Tg, ppm/°C
Flexipoxy 100 film adhesive (x–y plane)	1	−19.6	1.29	36.5
	2	20.0	4.19	36.6
Flexipoxy 100 film adhesive (z plane)	1	−15.3	291	46.6
	2	−18.7	319	47.3
Flexipoxy 100 paste adhesive	1	−22.2	124	41.6
	2	−23.0	127	41.9
Uralane 7760 paste adhesive	1	−77.6	132	43.5
	2	−78.7	129	39.6

Comments:

The thermal expansion data for the Flexipoxy 100 paste adhesive and the Uralane 7760 paste adhesive are typical of what we would expect of an elastomeric material. The CTE above Tg is about triple that of the CTE below Tg. However, the thermal expansion data of the film adhesive by contrast made me do a double take. The opposite of normal elastomeric behavior is observed. The CTE below Tg is perhaps 10 times that of the CTE above Tg. The presence of the woven glass cloth in the center of the film adhesive accounts for the anomaly. Below Tg, the polymer is hard and strong, and therefore has the properties of a reinforced plastic. However, above Tg, the polymer is soft and rubbery and has little contribution to the composite behavior. The glass cloth dominates and the composite exhibits negligible expansion.

5.5 RHEOLOGICAL ANALYSIS OF FLEXIPOXY 100 AND URALANE 7760 ADHESIVES

The dynamic shear modulus of the cured adhesives as a function of temperature was determined using a Rheometrics RDS 1700 tester. Uralane 7760 was tested in the temperature range −100°C to +100°C because of its low Tg. Flexipoxy 100 was tested between −50°C and +100°C. The frequency of the oscillations was 1 Hz and the temperature was raised at 3°C/min.

TABLE 5.4   Dynamic Shear Modulus at 50°C and 100°C

Adhesive	Run	Tg, °C	G' at 50°C, psi	G' at 100°C, psi
Flexipoxy 100 paste	1	−9	1400	1900
Flexipoxy 100 paste	2	−9	1600	1700
Flexipoxy 100 film	1	−3	Film delaminated	Film delaminated
Flexipoxy 100 film	2	−6	3300	1200
Uralane 7760 paste	1	−66	530	700
Uralane 7760 paste	2	−67	460	410

Comments:

Rheometrics was chosen to reduce the cost of testing. A plot of dynamic shear modulus versus temperature is obtained in a single run on a specimen. Traditionally, specimens would be made and tested at several temperatures to provide a modulus versus temperature plot.

However, the rheometrics values may be artificially high compared with static testing. Elastomers behave as viscoelastic materials and give different results depending upon rate of loading and rate of temperature change.

5.5.1 SUMMARY OF TASK 1

The stress analysts used the data that we provided to structurally evaluate the Flexipoxy 100 film or Uralane 7760-bonded flatpack configuration and they obtained acceptable results.

Additionally, we learned more about these new adhesives:

1. Flexipoxy 100 film adhesive is anisotropic due to the woven glass cloth reinforcement. It has a low CTE in the xy-plane (1–4 ppm/°C) and a markedly higher CTE in the z direction (300 ppm/°C). Its tensile strength is 3284 psi in the xy-plane and 553 psi in the z direction. The double lap shear strength at −34°C is at least 2100 psi and probably much higher.
2. Flexipoxy 100 paste adhesive is isotropic. It has a CTE of 125 ppm/°C, a tensile strength of 553 psi, and an ultimate elongation of 12%. It has a compressive strength of 1870 psi and a Poisson's ratio of 0.40.
3. Uralane 7760 paste adhesive is isotropic. It has a CTE of 131 ppm/°C, a tensile strength of 458 psi, and an ultimate elongation of 286%. It has a compressive strength of 1120 psi and a Poisson's ratio of 0.40. The double lap shear strength at −34°C is 320 psi.

5.2 TASK 2: SPACE-SIMULATED AGING METHODS FOR ELECTRONIC ASSEMBLY ADHESIVES

The NASA outgassing test (SP-R-0022) is a quick test to determine whether a given material will be suitable for a space environment. However, it provides data from a 24-h test at a single temperature in a space-like vacuum. The steering committee was interested in learning more about the expected life on orbit of the two materials and that required a larger test program. One of the TSD members of the steering committee was Dr. Tom Sutherland. He was primarily in charge of conducting this life prediction study.

5.2.1 MY PHYSICIST PROJECT PARTNER—DR. TOM SUTHERLAND

Tom Sutherland conducted a study of how the two new thermal transfer adhesives would fare in space over a spacecraft's lifetime. Tom actually ran this project, while my role was one of advising and assisting him. Our study was done through a technique called accelerated aging, that is, using testing of samples at elevated temperatures to simulate longer times at lower temperatures. Design engineering wanted the answers as soon

as possible, not after dozens of spacecraft were in orbit with potential un-
known problems caused by unforeseen problems with a new adhesive.

Tom Sutherland had been involved in all of the thermal transfer adhesive development activities, directed to the replacement of polysulfide. I
did not mention him previously because there were just too many names
in this team effort to mention them all. I liked working with Tom. We
seemed to be on the same wavelength. He was a huge asset to me in my
ATEP Materials Engineering course. He usually guest lectured for me in
two or three of the 20 lectures, and I could always count on him to help
me develop new lecture topics. Tom, who holds a Ph.D. in Physics, had
been a college professor for 10 years in Arizona before joining HAC. He
is a skillful and effective lecturer, and I learned a few tricks by observing
him in action. For example, he had a method for keeping the students from
falling asleep. The classes were held after working hours when folks were
already tired. The seats were like those in a movie theater, soft and com-
fortable. The lights were dimmed to make the projected slides more vivid.
All of the conditions for drifting off were there and some people couldn't
help falling asleep. Tom's technique was to occasionally get the attention
of one of the student in the class, saying "You there! In the yellow sweater,
What is your thinking about what I just said?" Nobody wants to be embar-
rassed, so they tried harder to stay awake.

5.2.2 THEORY UNDERLYING ACCELERATED AGING

In terrestrial applications, one important consideration is the presence of
oxygen, which may react with different polymeric materials and hasten
their degradation. This condition also exists for spacecraft electronic assembly materials during the time the spacecraft awaits launch. However,
once launched into orbit, oxygen, ozone, and other reactive gases are no
longer present and can be dismissed. Thermal degradation in the absence
of reactive gases is called "pyrolysis." The word derives from the Greek
"pyro" for fire and "lysis" for separating. Volatile fragments fly out of the
polymeric mass, and the phenomena can be monitored by continuously
measuring the weight of the specimen and plotted as mass loss versus
time. The thermogravimetric analyzer (TMA) is such a measuring device.

We can understand accelerated testing better if we consider the composition of our two adhesives. The alumina filler is about 70% of the weight

of the adhesives; fumed silica is also a small weight percentage. Neither of these ingredients will be affected by the temperatures under consideration. Essentially, all of the changes will be in the polymer phase of the adhesive. The polymers in these two adhesives consist of elastomeric, three-dimensional networks. Assuming that the composition of Hacthane 100 is nearly identical to that of Uralane 7760, we can discuss the polymeric part of the filleting adhesive. Hacthane 100 has a polymeric network composed of polybutadiene segments linked together by an aliphatic polyisocyanate (Desmodur N3200). Desmodur product data sheets list the functionality of N3200 as being between 2.8 and 3.6. In other words, three or more polybutadiene chains may be attached to each Desmodur N3200 molecule. The chemical structure between isocyanate groups in Desmodur N3200 is a string of methylene groups. In general, thermal stability is enhanced by increased cross-linking because severing one or two chains still leaves another to hold the network together and the broken links may even reattach if the ends remain in close proximity. There are essentially no non-reactive ingredients in Hacthane 100. We should expect good thermal stability from either Uralane 7760 or Hacthane 100.

Flexipoxy 100 is also a three-dimensional network polymer. Two different flexible epoxy resins would be linked by hexane diamine curing agent. Hexane diamine is a di-primary amine, which gives it a functionality of four. In other words, up to four epoxy resin chains may be linked to each hexane diamine molecule. So, there is a high level of cross-linking. Epon 871 was one of the flexible epoxy resins and DER 732 was the other. Epon 871 has an ester linkage in the molecule, whereas DER 732 had ether linkages. Flexipoxy 100 also has a non-reactive plasticizer (LHT-34 or LHT-28) in the formulation to make it weaker and more reworkable. The plasticizer has a polyoxypropylene backbone polymer similar to that in DER 732. Due to the fact that the plasticizer is high molecular weight and that it would have an affinity for the DER 732, we felt it would be non-migratory. Chain scission of the plasticizer due to high temperatures could result in volatile mass loss because it is not chemically tied into the three-dimensional polymer network.

The Arrhenius Equation is often quoted as the underlying basis for accelerated aging. Basically, it allows us to predict how chemical reaction rates change as we vary the temperature of the reaction. For many reactions, the rate doubles for every 10°C increase in temperature. We are interested in predicting out to 10 years. So, if we saw no significant damage to our test samples after 1 week exposure at 200°C, that would be equivalent to 10 years at our flatpack operating temperature of 93°C. Perhaps, 220°C is too severe and may fry the sample. Two weeks at 190°C is also equivalent to 10 years at 93°C. As is 4 weeks at 180°C or 8 weeks at 170°C and so forth.

Polymeric materials undergo mechanical changes over time. This phenomenon is called "viscoelastic behavior." For example, dimensional change in a test specimen under a constant load is called "creep." Stress relaxation is a similar phenomenon. An entire science of viscoelasticity exists. One of the more useful concepts arising from this discipline is the "time/temperature superposition theory." Like the Arrhenius equation, predictions of future behavior can be made by measuring mechanical changes in the test material. If you want to learn more about the Arrhenius equation or the time/temperature superposition theory, Wikipedia has done excellent articles on them and the Internet is at your fingertips.

5.2.3 PLANNING THE ACCELERATED AGING PROGRAM

First the operating environment for the thermal transfer adhesives was examined. It was determined that the adhesives must be able to operate effectively at a continuous 93.3°C (200°F) for 10 years. The following tests were selected to appraise longevity:

TABLE 5.4   Tests Chosen to Determine Long-Term Suitability in Space

Test	Significance
Volume resistivity	Electrical isolation of circuits
Thermal conductivity	Adequate heat transfer from components
Torsional shear strength	Retention of adhesion and thermal transfer ability
Flexibility	Resist thermal stresses
Dynamic shear modulus	Appraises stiffness across temperature range of interest
Glass transition temperature	Thermal range behavior
Vacuum isothermal mass loss	Fast test to appraise thermal stability vs. temperature
Continuous mass loss	Quantitative correlation to property changes during aging

FLATPACKS RUN HOTTER THAN FILLETED COMPONENTS

Based on the differences in thermal environments for flatpacks (Flexipoxy 100) versus filleted components (Uralane 7760), different aging tests regimes were selected:The thermal transfer adhesives under flatpacks saw higher continuous temperatures than the filleting adhesives used on diodes, capacitors, resisters, and so on. So our accelerated aging study was highly concerned with thermal degradation of Flexipoxy 100 adhesive, but not very concerned with thermal degradation of Uralane 7760 because it only saw mild temperatures. Our main concern with Uralane 7760 filleting adhesive was the mechanical stress of launching the satellite into orbit. We also want to be assured that it maintains its flexibility so that it does not stress and break delicate components. Table 5.6 describes the test regime for the two different adhesives.

TABLE 5.6   Test Program Overview for Flexipoxy 100 Adesive and Uralane 7760 Adhesive

Test	Test temperatures for Flexipoxy 100 adhesive, °C	Test temperatures for Uralane 7760 adhesive, °C
Volume resistivity	100, 125, 150, 225	
Thermal conductivity	150, 225	
Torsional shear strength	150, 225	
Flexibility		100, 125, 175, 225, 250
Dynamic shear modulus	100, 125, 150	
Glass transition temperature	100, 125, 150	
Vacuum isothermal	100, 125, 150	100, 150, 200, 250
TGA mass loss		
Continuous mass loss	100, 125, 150, 175	100, 125, 150, 175

5.2.4 PRELIMINARY TESTING

We conducted a quick screening test in order to get an idea of how the two adhesives would fare at elevated temperatues under vauum. The following table shows the percentage mass loss after 2 h at four different temperatures and under vacuum.

TABLE 5.7   Isothermal Temperature Exposure for 2 h under Vacuum (Thermogravimetric Analysis)

Temperature, °C	Mass loss, % Flexipoxy 100 adhesive	Mass loss, % Uralane 7760 adhesive
100	0.6	0.2
150	0.7	0.3
200	1.4	0.4
250	12.7	2.0

Uralane 7760 adhesive is obviously more thermally stable than Flexipoxy 100 adhesive. For either of them, thermal decomposition appears to be underway at 200°C. For temperatures below 150°C, both adhesives appear to be stable.

5.2.5   TEST RESULTS

Armed with the knowledge that Uralane 7760 is far more thermally stable than Flexipoxy 100 adhesive plus the fact that Flexipoxy 100 has the harder job of surviving 10 years on orbit at 93°C continuous service, our focus was concentrated on Flexipoxy 100 adhesive. Table 5.8 shows the results of monitoring its volume resistivity, thermal conductivity, glass transition temperature, and dynamic shear modulus during extended simulated and accelerated space aging.

TABLE 5.8   Results of Accelerated Space Aging on Flexipoxy 100 Film Adhesive

Property	Initial value	Aging	Result
Volume resistivity, ohm-cm	4.4×10^{11}	444 h at 100°C	No change
		321 h at 150°C	No change
Thermal conductivity, BTU/hr-ft-°F	0.22	321 h at 150°C	No change
		15 h at 225°C	Value halved
Glass transition temperature, ° C	−12.5	444 h at 100°C	Tg rose to +4
		321 h at 150°C	Tg rose to +7
Dynamic shear modulus, psi	8150 at 50°C	444 h at 100°C	Very slight increase
		150 h at 150°C	Very slight increase
		200 h at 125°C	3.6-fold increase
Torsion shear strength, in-lb	13.7	3 h at 225°C	Climbed to 50.8
		15 h at 225°C	Fell to 9.5

The volume resistivity remained constant and that is an important property for the electronic package. Thermal conductivity halved at 225°C indicating changes in the polymer phase of the adhesive. We had already established that thermal decomposition occurs at this temperature and that 225°C may just be too high for the purpose of accelerated testing. Modest increases in the glass transition temperature most likely are due to further cross-linking of the polymer. These cross-linking reactions may never occur at lower realistic service temperatures due to steric hindrance. The torsional shear data suggest more cross-linking which strengthened the polymer, followed by thermal degradation, which weakened it.

5.2.6 SUMMING UP THE LIFE PREDICTION STUDY FINDINGS

The life expectancy of three thermally conductive adhesives was determined at elevated temperatures in a simulated space environment. The three alumina-filled adhesives were Pro-Seal 727 (a polysulfide compound), Uralane 7760 (a polyurethane compound), and Flexipoxy 100 (an epoxy rubber compound). Quick comparative results were obtained on these three adhesives using isothermal gravimetric analysis. Uralane 7760 was the most stable, Flexipoxy 100 was a close second, but the polysulfide adhesive was a distant third. Despite its comparative instability, the polysulfide adhesive has flown on orbit successively for years in space satellites. The superiority of its two replacements is reassuring. More thorough testing was done using thermal vacuum aging and various tests.

5.2.7 PAPERS AND PRESENTATIONS

Dr. Tom Sutherland presented a paper, which he had co-authored with Ralph Hermansen, at the Fifth International SAMPE Electronics Conference in Los Angeles on June 20, 1991. The paper was titled, "Space-Simulated Aging for Electronic Assembly Adhesives."

5.3 TASK 3: PROBLEMS WITH CO-CURING POLYURETHANE AND EPOXY ADHESIVES

5.3.1 THE PROBLEM

Manufacturing ran into an unexpected problem. Some of the technicians applying components to PWB's had mixed uncured Uralane 7760 and Flexipoxy 100* adhesives so that the two liquids were physically in contact with each other. The material at their interface did not properly cure. The main concerns here are that the poorly cured, thermal transfer adhesive may not be sufficiently conductive to cool the electronic components and that there may be outgassing contaminants, which could cloud optics or solar cells. I was asked to help investigate the problem.

*Actually, the trade name was Ablefilm 5010K, but it was identical to Flexipoxy 100. Ablestik labs made the product to our provided formulation and procedure.

5.3.2 THE DIFFERING CHEMISTRY OF THE TWO ADHESIVES

Uralane 7760 and Flexipoxy 100 adhesives belong to two different families of polymers. Uralane 7760 is a polyurethane compound, whereas Flexipoxy 100 is an epoxy compound. Polyurethane polymers, like Uralane 7760, are the result of reacting polyols and polyisocyanates together. Epoxy polymers, like Flexipoxy 100, are the result of reacting epoxy resins and diamines together.

INSIDE THE MIXED ADHESIVES

If we were to thaw some frozen premix Uralane 7760 and thaw some frozen premix Flexipoxy 100 and mix them together, then we would have combined four different types of reactive ingredients together. Those reactive ingredient types are epoxy resins, diamines, polyols, and polyisocyanates. It only takes me a single glance at this list to spot a major problem: isocyanates and amines react with each other instantaneously! The product of their reaction is a bi-substituted urea. The resulting problem is diminishment of polyisocyanate needed to react with the polyols and reduction of diamine needed to cure the epoxy resins.

THE POLYISOCYANATE IS MOST DIMINISHED

Assume that we have thoroughly mixed equal quantities of Uralane 7760 and Flexipoxy 100 pastes. The isocyanate ingredient is close to 3% of the Uralane 7760 paste and the diamine is close to 3% of the Flexipoxy 100 paste. However, their equivalent weights are quite unequal. The isocyanate has an isocyanate equivalent weight of 116 and the diamine an amine equivalent weight of 58. So after the fast reaction between these two reactants, all of the isocyanate would be consumed forming a urea compound, whereas only half of the diamine would be consumed. Half of the original diamine is still available to react with epoxy resin.

5.3.3 EXPERIMENTAL DESIGN

Two different experimental plans were conducted:

EXPERIMENTAL PLAN 1

Test specimens according to the following experimental plan:

TABLE 5.9 Weight Ratio of Two Adhesive Mixture Trials

Mixture	A	B	C	D	E
Uralane 7760 paste	15	30	50	70	85
Flexipoxy 100 paste	85	70	50	30	15

Each mixture (A–E) was prepared in three different ways:

M1 = thoroughly mixed, M2 = minimal mixing, and M3 = just touching.
Specimens were cured for 1 h at 200°F and then divided into three equally sized groups.
Group I = no postcure, Group II = 24 h at 200°F postcure, and Group III = 48 h at 200°F postcure.
For each specimen, the firmness was observed and unreacted material was extracted using chloroform. The extracts were subjected to Fourier transform infrared spectroscopy (FTIR) analysis to identify the unreacted ingredient.

Note: FTIR is short for Fourier transform infrared spectroscopy, which is a technique used to obtain an infrared spectrum of absorption of a gas, liquid, or solid. Rather than shining a monochromatic beam of light at a sample, this method shines a beam containing many frequencies at once and measures how much of that beam was absorbed by the sample. Next, the beam is modified to contain a different set of frequencies. This process is repeated many times. A computer program then works out what the absorption is at each frequency. FTIR analysis is an extremely useful tool to a formulator because ingredients can be identified and reactions can be tracked. Moreover, the FTIR spectra of an ingredient are like a fingerprint.

EXPERIMENTAL PLAN 2:

Three different materials were tested for thermal conductivity and NASA outgassing:

A. The current batch of Flexipoxy equivalent made by Ablestik.
B. The qualification batch of Flexipoxy equivalent made by Ablestik.
C. The current Ablestik Flexipoxy buttered with Uralane 7760.

5.3.4   TEST RESULTS

RESULTS FROM PLAN 1

Firmness

(A) With the exception of Trial A (85% Flexipoxy 100 and 15% Uralane 7760), all co-mixed trials were noticeably softer than the pure compounds.
(B) M1 (thoroughly mixed) was softer than M2 (minimal mixing) was softer than M3 (just touching).

FTIR analysis of extracts

FTIR scans were run on each of the reactive ingredients to assist in the identification. Epon 871 has a characteristic peak at 1725–1740 cm^{-1} and Desmodur N-3200 has a characteristic isocyanate peak at 2272 cm^{-1}. Of the reactive ingredients, it is the only one having a peak at 2272 cm^{-1}.

General comment for all levels of mixing:

(A) As predicted, there was no indication of unreacted isocyanate in any of the extracts for trials A through E, minimal cure, and any level of mixing (M1–M3). The same can be said for the absence of hexane diamine in the extract.

(B) Postcuring caused more of the epoxy resin to react as indicated by the smaller ester peak indicative of Epon 871.

For well-mixed M1 and poorly mixed M2 trials:

(A) The amount of PolyBd polyol in the extract increases as the amount of Uralane 7760 in the mixture increases.

(B) The amount of Epon 871 epoxy resin in the extract increases as the amount of Flexipoxy 100 in the mixture increases.

(C) C. Postcuring of 24 h reduced the epoxy resin in the extract. After 48 h of postcuring, the extract seemed to be mainly polyol. Both PolyBd R45HT in Uralane 7760 and LHT-34 in Flexipoxy are polyols.

RESULTS FROM PLAN 2

The thermal conductivity results are presented in the following tables:

TABLE 5.10 Thermal Conductivity Test Results (BTU/h-ft-°F)

Sample	C-Matic results	Colora results
Flexipoxy 100 from Current Ablestik batch	0.21, 0.19, 0.15	0.15, 0.17, 0.16
Flexipoxy 100 from Ablestik qualification batch	0.18, 0.15, 0.17	0.19, 0.18
Flexipoxy 100 (Ablestik) with Uralane 7760	0.24, 0.25	0.24, 0.28, 0.25

Outgassing results

NASA outgassing tests were run on all of the same samples. All of them passed, including the sample of mixed Uralane 7760 and Flexipoxy 100.

5.3.5   DISCUSSION OF THE TEST RESULTS

THE MIXED ADHESIVE ANALYSIS

The firmness observations of mixed adhesives being softer than expected tells us that mixing Uralane 7760 and Flexipoxy 100 causes an under-cured condition. Our prediction that the isocyanate ingredient would be consumed by diamine was confirmed by FTIR analysis of the extracts. The hexane diamine is also partially consumed in this reaction to form a urea linkage.

The thoroughly mixed samples represent the extreme condition, which is unlikely to ever occur during a buttering on of Uralane 7760 paste to Flexipoxy film adhesive. Yet, it provides valuable information about the worst case scenario and if these samples are deemed safe to use, then all lesser cases are acceptable too.

The thermal conductivity results were lower than expected from previous testing. However, the mixed adhesives samples had higher thermal conductivity than the control specimens. The hardware should be safe to use. Moreover, the NASA outgassing results tell us everything passed, including the mixed Uralane 7760/Flexipoxy 100 samples.

THE EFFECT OF POSTCURING TO ADVANCE THE STATE OF CURE

Extract analysis also showed that postcuring diminished the amount of epoxy resin indicating that more of it was tied up as high-molecular-weight rubber. After 48 h of postcuring, the extract seemed to be predominately polyol.

Flexipoxy 100 is capable of converting to a rubbery state even if the hexane diamine is half of the theoretical amount needed. In order to understand this, we need to consider the functionality of the ingredients. Both epoxy resins, Epon 871 and DER 736, are diepoxides and are therefore difunctional. However, hexane diamine has four active hydrogens and is therefore tetrafunctional. In other words, hexane diamine can link to four epoxide groups, which themselves may be the terminal ends of growing chains. If the hexane diamine only links two epoxide groups, a polymer can still be formed.

Now, here is why a heat cure can continue polymerization: Hexane diamine has two primary amine groups which are very reactive. However, after they each react with an epoxide group, the two remaining active hydrogens are secondary amines, which are less reactive. Moreover, they are now sterically hindered. Elevated temperatures increase molecular motion and increase the likelihood of their reaction with an epoxide group.

There are also active hydrogens on the urea groups formed from the reaction of the isocyanate in Uralane 7760 and the diamine in Flexipoxy 100. Elevated temperatures increase the likelihood of an epoxy group linking to the urea group forming a cross-link. This is another mechanism for the free epoxides to be tied up.

The least likely ingredients to become part of the growing polymers are the polyols. Although we have seen that PolyBd R45HT can react with itself through the double bonds in its chain, this reaction has been observed during heat aging of cured Uralane 7760 and it begins at the air surfaces of the sample and slowly progresses inward with time at temperature. LHT-34 is a high-molecular-weight triol and has an affinity to the DER 736 epoxy resin in Flexipoxy 100 due to their polyoxypropylene chains.

5.3.6 SUMMARY OF TASK 3

Mixed Uralane 7760 and Flexipoxy 100 caused a noticeably under-cured state on production hardware. The adhesives were not intended to be cured together and in contact with one another. The isocyanate in the Uralane 7760 compound reacts vigorously with the diamine in the Flexipoxy 100 compound, leaving insufficient isocyanate to cure the polyol in Uralane 7760 and insufficient diamine to properly cure the Flexipoxy 100. Intimate mixing of the two adhesives in a range of ratios shows that an under-cured, softer firmness exists which is worst for mixtures with mainly Uralane 7760. Chloroform extraction of uncured material from these samples shows that epoxy resin and polyols are mostly found and not isocyanate or diamine. Postcuring ties up more of the epoxy resins into the polymer. In cases of minimally mixed Uralane 7760/ Flexipoxy 100 samples, the samples were firmer and less under-cured.

Testing of thermal conductivity on Uralane 7760 buttered on Flexipoxy 100 samples showed it had at least equal thermal conductivity to uncontaminated Flexipoxy 100. Moreover, the sample passed NASA outgassing.

KEYWORDS

- Uralane 7760
- Flexipoxy 100
- CTE
- epoxy polymers
- DER 736

CHAPTER 6

SUPERIOR THERMAL TRANSFER ADHESIVE

CONTENTS

6.1   THE NEED FOR A BETTER THERMAL TRANSFER ADHESIVE APPROACH

The thermal transfer adhesive approach in 1990 consisted of two different adhesives: Uralane 7760 paste adhesive or its equivalent, Hacthane 100, for fillet bonding of discrete components and Flexipoxy 100 film adhesive for bonding of flat-bodied components. This approach was a vast improvement over the previously used polysulfide adhesives, which had unacceptably bad outgassing performance. Nonetheless, there were a few problems with the 1990 system, namely,

(1) MUX circuits: Flexipoxy 100 needs better dielectric properties. There has been micro-amp leakage at elevated temperatures (93°C). This problem did not exist with the previous polysulfide adhesive.
(2) The co-curing of Flexipoxy 100 adjacent to Uralane 7760 or Hacthane 100 inhibits proper cure. This was discussed in the previous chapter.
(3) A lower glass transition temperature for the film adhesive would assure against hardening of the adhesive with time in service.

Our formulating knowledge of flexible epoxies had advanced considerably since the development of Flexipoxy 100, so we proposed a development plan to formulate an advanced version of Flexipoxy 100, which had better dielectric properties, lower glass transition, and so on, while still meeting all of the other requirements of the application. Robert B. Mitsuhasihi, manager of the Space Electronics Department, funded our development effort and monitored our progress closely.

6.2   A NEW FORMULATION EFFORT

6.2.1   PROJECT GOALS

The new adhesive should meet all of the current requirements, but also have a volume resistivity of 10^{12} ohm-cm at 200°F in order to solve the micro-amp current leakage problem seen with Flexipoxy 100 adhesive on MUX circuits. It is also desirable and feasible to lower the glass transition temperature as much as possible so that the adhesive has rubbery properties over a wide temperature range.

6.2.2 PROJECT SCHEDULE

On September 17, 1990, the project to develop the next-generation thermal transfer adhesive was deemed important enough by the Space and Communication Group to appoint a steering committee to oversee it. I informed Steve Lau of the higher visibility and the need to be responsive to action items assigned by the committee. Bob Mitsuhashi is the committee head.

The overall project is divided into phases as follows:

    Phase 1   Development of new Flexipoxy adhesive
    Phase 2   Full qualification
    Phase 3   Revision of specifications
    Phase 4   Technology transfer to Ablestik Laboratories
    Phase 5   Implementation

6.3 APPROACH

6.3.1 NEW INGREDIENTS MAKE PROGRESS POSSIBLE

Our discovery of the flexible Cardolite epoxy resins and Witco's DP 3680 curative opened the door to Flexipoxy formulations with superior dielectric properties and extended temperature range where the adhesive is still flexible. The chemical structures of Cardolite NC-513, Cardolite NC-514, and DP-3680 are shown in the figures below:

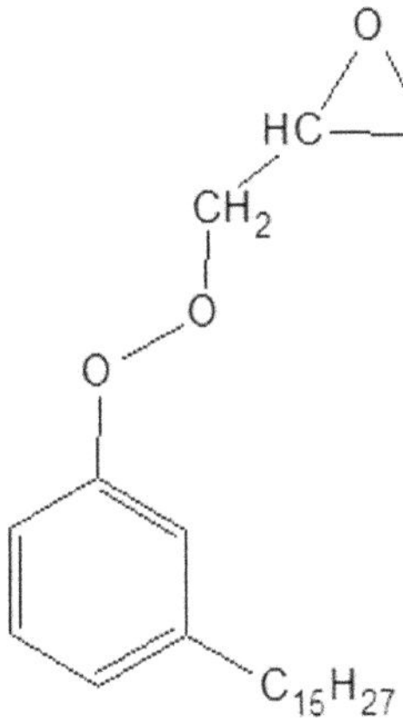

FIGURE 6.1   Cardolite NC-513 monoepoxide.

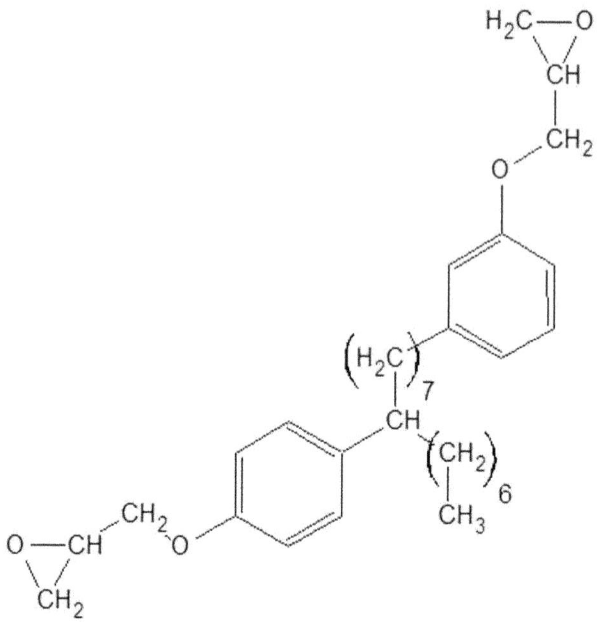

FIGURE 6.2   Cardolite NC-514 epoxy resin.

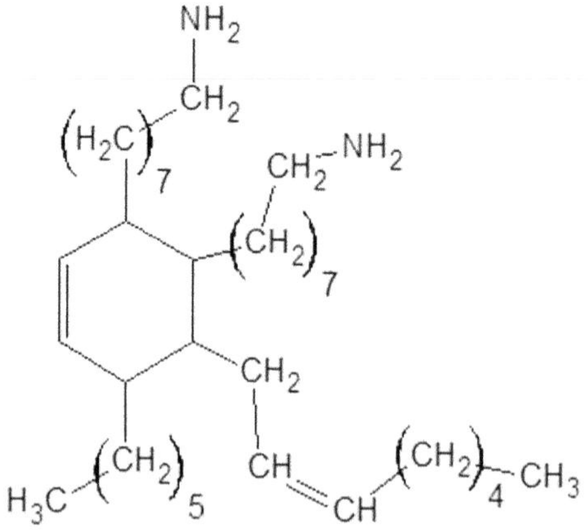

FIGURE 6.3   DP-3680 diamine.

6.3.2 OUR APPROACH FOR FASTER FORMULATION DEVELOPMENT

Although the end goal is to develop a thermal transfer adhesive, it was possible to get more experimental data quicker by only examining the polymeric part of the thermal transfer adhesive and ignore the filler portion. It is far more time-consuming to prepare batches of filled compound in the Ross mixer than to just stir up the liquid ingredients in a beaker. Dielectric properties, for example, do not change significantly when the filler is added. Glass transition temperature is also virtually the same, filled or unfilled. NASA outgassing is significantly affected by the filler because the filler adds inert material to the adhesive. Yet, the weight loss and captured condensable matter all derive from the polymeric portion of the adhesive. One can mathematically calculate from unfilled adhesive outgassing data what the filled adhesive's outgassing properties will be.

6.4 FORMULATING EXPERIMENTS TOWARD A SUPERIOR FILM ADHESIVE

6.4.1 PLASTICIZER EVALUATION EXPERIMENT

The objective of this experiment was to evaluate better plasticizers than LHT-34 or LHT-28 as was used in Flexipoxy 100. The polyoxypropylene chain segment in these two triols is hydrophilic, and we have seen that hydrophobic chain segments tend to improve dielectric properties. Using Flexipoxy 160 as an unplasticized control, trial Flexipoxy 164 had the polyol PolyBd R45HT as a plasticizer, trial Flexipoxy 171 had PolyBd R20LM as a plasticizer, and trial Flexipoxy 170 had castor oil as the plasticizer.

In the table below, both the candidate formulations and their resultant properties as cured adhesives are presented.

TABLE 6.1 Flexipoxy Adhesive Formulations with Different Plasticizers and Their Properties

Ingredient	Flexipoxy 160	Flexipoxy 164	Flexipoxy 170	Flexipoxy 171
Cardolite NC-514	90 pbw	90 pbw	90 pbw	90 pbw
Cardolite NC-513	10 pbw	10 pbw	10 pbw	10 pbw
DP-3680	47.2 pbw	47.2 pbw	47.2 pbw	47.2 pbw
PolyBd R45HT		30 pbw		
PolyBd R20LM				30 pbw
Castor oil			30 pbw	
Volume resistivity at 75°F	1.5×10^{15} ohm-cm	1.1×10^{14} ohm-cm	8.2×10^{13} ohm-cm	1.6×10^{14} ohm-cm
Volume resistivity at 200°F	4.2×10^{11} ohm-cm	4.1×10^{11} ohm-cm	3.3×10^{11} ohm-cm	4.0×10^{11} ohm-cm
Glass transition temperature	+4.3°C	−2°C	−7.8°C	−19.7°C
Outgassing TML-WVR	1.26%	0.71%	0.72%	1.21%
Outgassing CVCM	0.27%	0.08%	0.14%	0.29%

The volume resistivities of all these new adhesives are at least two decades better than Flexipoxy 100 adhesive at both at room temperature and 200°F. The volume resistivities of the new adhesives at 200°F seem to be independent of plasticizer. All of the values are very similar.

I am a little dubious about these glass transition temperature results. R45HT and R20LM are similar polybutadiene polyols and I would have expected them to lower the T_g about the same amount. Perhaps the samples were mislabeled and Flexipoxy 164 actually contains castor oil and 170 contains R45HT. That would make the results closer to my expectations.

Outgassing results are interesting in that the addition of plasticizer did not worsen the results. Apparently, these polyols are chemically tying into the polymeric matrix. FTIR analysis of the volatile condensible matter showed it to be a monoepoxide diluent. The Cardolite resins are the likely source of the contaminant. In fact, NC-513 is a monoepoxide diluent.

6.4.2 FLEXIPOXY 171 EVALUATED WITH ALUMINA FILLER

Flexipoxy 171 was the most promising candidate as of October 1990. It was made into a thermal transfer adhesive by adding alumina powder using the Ross dual planetary mixer, and then evaluated along with the unfilled version. The formulations are as follows:

TABLE 6.2   Flexipoxy 171 Filled and Unfilled Paste Adhesive Formulations

Ingredient	Unfilled version	Filled version
Cardolite NC-514	50.79	11.7
Cardolite NC-513	5.64	1.3
Atochem R20-LM	16.93	3.9
Witco DP-3680	26.64	6.1
Cab-O-Sil	none	2.0
325 mesh alumina	none	75.0
Total	100.00	100.00

The properties are as follows:

TABLE 6.3   Properties of Flexipoxy 171 Filled and Unfilled Adhesives

Property	Unfilled version	Filled Version
Volume resistivity at 75°F	1.6×10^{14} ohm-cm	3.1×10^{14} ohm-cm
Volume resistivity at 200°F	4.1×10^{11} ohm-cm	No test
Glass transition temperature	−19.7°C	−17.7°C
Outgassing: TML-WVR	1.21%	No test
Outgassing: CVCM	0.29%	No test
Durometer	No test	93 Shore A
Cure-to-set time at 200°F	No test	18–24 min

Although far superior to Flexipoxy 100, Flexipoxy 171 is deemed too firm, and has marginal CVCM.

6.4.3 NEW CARDOLITE NC-547 EPOXY RESIN EXTENDS ADHESIVE'S FLEXIBILITY TO LOWER TEMPERATURES

Cardolite NC-547 became an epoxy resin of great interest in October 1990. It imparted extraordinarily low-temperature flexibility (i.e.,°−55°C)and in doing so made its Flexipoxy adhesives similar to Hacthane 100 adhesive.

TABLE 6.4 Formulation of Flexipoxy 175F Thermal Transfer Adhesive

Ingredient	Weight %
Cardolite NC-547	13.0
Witco DP-3680	5.1
Cab-O-Sil	0–1.9
325 mesh alumina	75–76.9

The properties of Flexipoxy 175F Paste Adhesive are compared with our current thermal transfer adhesive, Flexipoxy 100, in the following table:

TABLE 6.5 Comparison of Flexipoxy 175F Paste Adhesive versus Flexipoxy 100 Paste Adhesive

Property	HMS 2308 Requirement	Flexipoxy 100 paste	Flexipoxy 175F paste
Work-Life	>2 h	>2 h	> 2 h
Durometer, Shore D	38–58	49	45
Freezer storage life	>3 months	>3 months	>3 months
Color	Off-white	Off-white	Tan
Specific gravity	2.0–2.2	2.1	2.3
Solvent resistance	<7% weight gain	4.5%	No test
Volume resistivity	$> 1 \times 10^{11}$ ohm-cm	$1-3 \times 10^{11}$ ohm-cm	3×10^{14} ohm-cm
Outgassing:			
TML-WVR	<1.0%	0.6%	0.41%
CVCM	<0.1%	0.04%	0.12%
Thermal Conductivity,			
BTU/h-ft-°F	>0.25	0.30–0.35	0.48–0.55 calculated
Glass Trans. Temp.	<4°C	−5 to −12°C	−55°C
CTE>T_g	<171 ppm/°C	150 ppm/°C	115 ppm/°C
CTE<T_g	< 52 ppm/°C	40 ppm/°C	45 ppm/°C
Lap shear strength	100–700 psi	400–500 psi	410 psi

Flexipoxy 175F is far superior to Flexipoxy 100 in volume resistivity, thermal conductivity, and glass transition temperature. However, its CVCM is not acceptable. Formulation adjustments are needed to get the CVCM under 0.1% max requirement.

Flexipoxy 175F was also used to make a thermal transfer film adhesive. Its properties are compared with Flexipoxy 100 film adhesive in the Table 6.6.

TABLE 6.6 Comparison of Flexipoxy 175F Film Adhesive versus Flexipoxy 100 Film Adhesive

Property	HMS 2308 requirement	Flexipoxy 100 paste	Flexipoxy 175F paste
Construction	Two step	Two step	One step See Note 1
Film thickness	12–18 mils	14-18 mils	18 mils
Work-life	>2 h	>2 h	>2 h
Lap shear strength	100–700 psi	300–500 psi	400 psi, see Note 2
Freezer storage life	>3 months	>8 months	TBD
Color	Off-white	Off-white	Tan
Volume resistivity	$> 1 \times 10^{11}$ ohm-cm	$1\text{–}3 \times 10^{11}$ ohm-cm	3×10^{14} ohm-cm
Dielectric constant	10.0 max	8.8-9.8	No test
Dissipation factor	0.150 max	0.50–0.11	No test
Dielectric strength	>350 V/mil	500–750 V/mil	No test
Outgassing:			Estimated
TML-WVR	<1.0%	0.6%	About 0.41%
CVCM	<0.1%	0.04%	About 0.12%
Thermal conductivity, BTU/h-ft-°F	>0.25	0.30–0.35	0.48–0.55 calculated
Lap shear strength	100–700 psi	300–500 psi	About 400 psi See Note 2

Note 1: Two-step process is made unnecessary due to the high-volume resistivity of Flexipoxy 175F.

Note 2: Estimated from paste adhesive lap shear strength.

6.5 FORMULATING EXPERIMENTS TOWARD BOTH SUPERIOR FILLETING ADHESIVE AND SUPERIOR FILM ADHESIVE

6.5.1 CHANGE IN OBJECTIVES

By January 28, 1991, thanks to Cardolite NC-547, our new flexible epoxy formulations had both improved volume resistivity and markedly lower glass transition temperatures. We were approaching the superior properties already in our filleting adhesive, Hacthane 100. The Steering Committee decided to broaden the goal of the development effort from just improving the film adhesive to also replacing the current filleting adhesive with the new adhesive. With both filleting and film adhesive being of the same chemistry, problems of incompatibility would disappear.

However, there was still one big problem to be solved. The outgassing results for the new adhesive were not acceptable. We kept failing the 0.01% maximum requirement for CVCM.

6.5.2 THE VOLUME RESISTIVITY GOAL IS CHALLENGING

Steve Lau was finding it difficult to formulate a Flexipoxy compound with a volume resistivity of at least 1×10^{12} ohm-cm at a test temperature of 200°F. He wondered if it was even possible. He ran three simple formulations of a single resin cured with DP-3680 through the tests, while including Flexipoxy 100 and Flexipoxy 175 for comparison. The simple formulation resins were Epon 872, NC-514, and NC-547. Flexipoxy 175 is also based on NC-547 and DP3680, but also contains alumina powder and silica. In late October 1990, we reported the following data:

TABLE 6.7   Volume Resistivity of Various Flexipoxy Compounds as a Function of Test Temperature, in Ohm-cm

Test temperature, °F	Flexipoxy 100 paste	Epon 872/ DP3680	NC-514/ DP-3680	NC-547/ DP-3680	Flexipoxy 175
75	1.6×10^{11}	1.9×10^{16}	5.3×10^{15}	1.2×10^{14}	3.6×10^{14}
120	3.6×10^{10}	1.3×10^{13}	2.8×10^{13}	1.3×10^{13}	2.0×10^{13}
160	8.5×10^{9}	6.4×10^{11}	1.8×10^{12}	1.8×10^{12}	1.8×10^{12}
200	4.8×10^{9}	1.3×10^{11}	3.5×10^{11}	2.7×10^{11}	2.7×10^{11}

Comments:

None of the trials attained the desired 1×10^{12} ohm-cm at a test temperature of 200°F. Surprisingly, even those with quite high room temperature values still ended up in the 10^{11} ohm-cm range. Yet, all of the four new trial compounds had volume resistivities that were almost two decades better than the value attained by the Flexipoxy 100 adhesive. Perhaps this might be a big enough improvement to solve the micro-amp leakage problem.

6.5.3 VCM REDUCTION SERIES #1

In January 1991, the obstacle to accepting Flexipoxy 175F2, which is the alumina-filled adhesive based on NC-547 and DP-3680, was its high CVCM in the outgassing test. It was pointed out in the Steering Committee meeting that the film adhesive would see several hours of postcure during subsequent operations and perhaps that would help pass the CVCM requirement. In order to verify this premise, Flexipoxy 175F2 was subjected to various levels of vacuum stripping and postcuring in an attempt to lower the CVCM of the adhesive. The results are shown in the following table:

TABLE 6.8   Effect of Special Processing on the Outgassing of Flexipoxy 175F2 Adhesive

Condition or property	Control	Trial 1	Trial 2	Trial 3	Trial 4	Trial 5
Vacuum stripping at 95°C and <200 torr	None	1 h	4 h	8 h	16 h	24 h
200°F cure	1 h	1 h	1 h	1 h	1 h	1 h
200°F postcure	None	None	4 h	8 h	16 h	24 h
Outgassing*						
TML, %	0.41	0.37	0.50	0.50	0.50	0.41
CVCM, %	0.12	0.09	0.13	0.14	0.11	0.12

*Based upon 75% filled paste.

Strangely, neither vacuum stripping nor postcuring significantly improved the CVCM values. Perhaps, 200°F is not a high enough temperature. The outgassing test exposes the specimens to 24 h at about 250°F.

6.5.4   VCM REDUCTION SERIES #2

Continuing in our effort to reduce the outgassing of the formulation to an acceptable level, we reduced the Cardolite NC-547 level by substituting other flexible resins for part of it. The formulations tested are shown in the Table 6.9: Note that the first five candidates comprise a series, in which Flexipoxy 187 and 163 are the masterbatches and Flexipoxy 207,208, and 209 are intermediates made by mixing of the two masterbatches.

TABLE 6.9   Formulations of VCM Reduction Series #2

Ingredient	Flexipoxy 187	Flexipoxy 207	Flexipoxy 208	Flexipoxy 209	Flexipoxy 163
NC-547, pbw	72	54	36	18	0
NC-514, pbw	0	18	36	54	72
NC-513, pbw	8	8	8	8	8
R45HT, pbw	20	20	20	20	20
DP-3680, pbw	23.2	26.8	30.5	34.1	37.7

The test results for VCM reduction series #2 are shown in the table below: Glass transition temperature was obtained using a thermal differential analyzer.

Here we have a beautiful example of getting predictable intermediate values when mixing two masterbatches in different proportions. Here are some of those patterns:

1.   Cardolite NC-514 produces higher volume resistivity than NC-547, and we can see the values increase as we read left to right.
2.   Cardolite NC-514 produces better outgassing results than NC-547, and we can also see that as we read across.
3.   Cardolite NC-547 produces a lower Tg than NC-514, and that is also obvious as we read across from left to right.
4.   Cardolite NC-547 produces a lower durometer than NC-514, which is again obvious.

I tend to put more faith in the execution of an experiment where the resulting test values lie on a smooth curve. When test data are predictable and logical, that indicates to me that the formulation ingredient weighings and the testing were done precisely.

TABLE 6.10 Properties of VCM Reduction Series #2 Candidates

Property	Flexipoxy 187	Flexipoxy 207	Flexipoxy 208	Flexipoxy 209	Flexipoxy 163
Vol. Res., ohm-cm					
At 75°F	3.5×10^{13}	6.7×10^{13}	8.1×10^{13}	1.2×10^{14}	1.5×10^{14}
At 200°F	9.1×10^{10}	1.1×10^{11}	1.3×10^{11}	1.6×10^{11}	1.5×10^{11}
Outgassing					
TML,%	0.96	0.81	0.75	0.70	0.87
CVCM,%	0.23	0.17	0.11	0.04	0.09
Calculated outgassing*					
TML,%	0.29	0.24	0.23	0.21	0.26
CVCM,%	0.07	0.05	0.03	0.01	0.03
Glass transition temperature, °C	−59.9	−38.6	−26.7	−12.1	−2.1
Durometer	43 A	53 A	62 A	68 A	78 A

*Projected outgassing results for 70% filled film adhesive based upon unfilled paste values in the row above. The filler is inert and does not outgas.

6.5.5 VCM REDUCTION SERIES #3

Continuing our attempt to reduce VCM (Volatile Condensable Matter), we looked at the effect of replacing NC-547 resin in part with another epoxy resin and also the effect of adding PolyBd R45HT polyol to the formulations in different amounts. Table 6.11 shows the different formulation variations.

TABLE 6.11 VCM Reduction Series #3: Flexipoxy Trial Formulations

Formulation	223	224	225	226	227	228	229	187	230
NC-547*	95	95	95	92.5	92.5	92.5	90	90	90
NC-513	5	5	5	7.5	7.5	7.5	10	10	10
PolyBd R45HT	30	25	20	30	25	20	30	25	20
DP-3680	28.7	28.7	28.7	28.8	28.8	28.8	29.0	29.0	29.0

*Ingredients in parts by weight.

The test results for the trial formulations are of VCM reduction series # 3 presented in Table 6.12. Volume resistivity, NASA outgassing, and Shore A durometer are the properties of key interest.

TABLE 6.12 VCM Reduction Series #3: Volume Resistivity, Outgassing, and Durometer Results

Trial	Volume resistivity, ohm-cm at 75°F	Volume resistivity, Ohm-cm at 200°F	Outgassing TML, %	Outgassing CVCM, %	Shore A Durometer
Flexipoxy 223	5.8×10^{13}	7.9×10^{10}	1.19	0.38	50
Flexipoxy 224	7.4×10^{13}	8.8×10^{10}	1.23	0.34	53
Flexipoxy 225	8.7×10^{13}	8.6×10^{10}	1.06	0.24	55
Flexipoxy 226	5.0×10^{13}	7.1×10^{10}	1.11	0.30	49
Flexipoxy 227	6.2×10^{13}	8.3×10^{10}	1.18	0.39	52
Flexipoxy 228	6.7×10^{13}	1.1×10^{11}	1.14	0.35	53
Flexipoxy 229	4.0×10^{13}	6.3×10^{10}	1.03	0.28	48
Flexipoxy 187	5.8×10^{13}	4.6×10^{10}	0.96	0.23	49
Flexipoxy 230	5.8×10^{13}	6.1×10^{10}	1.28	0.50	48

Comments:

Volume resistivity: The goal for volume resistivity at 200°F is a value greater than 1×10^{12} ohm-cm. Flexipoxy 228 had the best results of the series, but it is short of the goal.

CVCM: These values would be reduced to about one-third of the reported values in the filled formulations, where 70% of the weight of the formulation is inert alumina powder. The TMLs would easily pass the 1.0% maximum requirement, but the estimated CVCMs for the filled compounds are marginal.

Durometer: All of the trials are essentially the same durometer.

The results overall are disappointing because not many clear trends are evident. It seems that increasing NC-513 amounts tend to lower volume resistivities and slightly lower durometers. The results may reflect a complex relationship between these ingredients and the state of cure of the adhesive. The monoepoxide, NC-513, reduces the degree of cross-linking which is possible. The PolyBd R45HT may also be tying into the growing polymer.

6.5.6   VCM REDUCTION SERIES #4

Still continuing our attempt to formulate an adhesive which passes the NASA outgassing requirements, we examined the ratio of reactive groups of the basic formulation. Theoretically, there should be one amine active hydrogen equivalent for each epoxide equivalent. However, the ideal ratio for best outgassing result might be different than theoretical.

In the following experiment, the amine hydrogen to epoxide index was varied from 0.7 to 1.3 in increments of 0.1 by increasing the parts-by-weight of DP-3680 while holding the Cardolite NC-547 constant at 100.0 pbw. One additional formulation was included which utilized hexane diamine instead of DP-3680 for comparison's sake. The following table shows the formulations, which were tested:

TABLE 6.13   VCM Reduction Series #4; Flexipoxy Trial Formulations

Ingredient	231	232	233	175	235	236	237	234
Amine H/epoxide index	0.7	0.8	0.9	1.0	1.1	1.2	1.3	1.0
pbw NC-547	100.0	100.0	100.0	100.0	100.0	100.0	100.0	100.0
pbw DP-3680	19.8	22.7	25.2	28.3	31.2	34.0	36.8	0.0
pbw hexane diamine	0.0	0.0	0.0	0.0	0.0	0.0	0.0	4.8

The formulation series was tested for volume resistivity at both 75°F and 200°F, outgassing, and Shore A durometer. The results are shown in the Table 6.14.

TABLE 6.14   Properties of VCM Series 4 Flexipoxy Trials

Flexipoxy ID	Volume resistivity, ohm-cm @ 75°F	Volume resistivity, ohm-cm @ 200°F	TML, %	CVCM, %	Shore A durometer
231	3.2×10^{14}	1.4×10^{11}	1.04	0.34	69
232	3.1×10^{14}	1.6×10^{11}	0.97	0.29	71
233	2.6×10^{14}	1.4×10^{11}	1.04	0.28	71
175	1.7×10^{14}	1.2×10^{11}	1.27	0.46	69
235	1.2×10^{13}	9.7×10^{10}	1.20	0.30	65
236	8.4×10^{13}	7.6×10^{10}	1.09	0.21	61
237	6.6×10^{13}	7.8×10^{10}	1.13	0.18	60
234	3.2×10^{14}	8.4×10^{10}	1.23	0.41	80

The volume resistivities of Flexipoxy 231–233 are higher than what we obtained in VCM reduction series #3. Apparently, volume resistivity is highest when NC-547 is the most dominant ingredient in the formulation. Unfortunately, CVCM is at its worst when that is true.

Volume resistivity trends: The volume resistivity data indicate that a slightly lower amine index yields higher volume resistivity. This may be the ideal reactive ingredient ratio and it may be that the NC-547 ingredient raises volume resistivity more than the DP-3680 ingredient.

The outgassing results are this series indicates that higher levels of DP-3680 improve CVCM results. This may be true because all of the epoxy groups have found an amine hydrogen to react with. Especially the monoepoxides need to be tied up so they are not part of the CVCM residue.

The Shore A durometer of the series (samples 231, 232, 233, 175, 235, 236, and 237, but not 234) is the classical curve generated when one goes off-ratio from 1:1. The more cross-linking, the flatter the curve generated when these data are plotted. The softening indicated by samples 175 >235 > 236 > 237 means that we have an ever-increasing excess of amine curative.

Formulations Flexipoxy 175 and Flexipoxy 234 should be directly compared to learn about the effects of the curative in the formulation. Both trials are at the theoretical 1:1 amine H to epoxide ratio. As expected, the hexane diamine compound (Flexipoxy 234) was harder than the DP3680 compound (Flexipoxy 175). Both of them had the highest CVCMs. The DP-3680 compound had better volume resistivity at 200°F.

6.5.7 ELECTRICAL PROPERTIES IMPROVEMENT SERIES

Bob Mitsuhashi of the Steering Committee wanted us to focus more on improving the volume resistivity of the adhesive. This series had the goal of improving volume resistivity by formulation modification. The formulations for the four candidate adhesives are shown in the Table 6.15.

TABLE 6.15 Flexipoxy Formulation Trials for Electrical Improvement Series

Ingredients	Flexipoxy 225.9	Flexipoxy 207	Flexipoxy 208	Flexipoxy 1100
Cardolite NC-547	65.15%	42.58%	27.59%	68.18%
Cardolite NC-514		14.20%	27.59%	
Cardolite NC-513	3.43%	6.31%	6.13%	
DDI 1410 isocyanate				0.69%
PolyBd R45HT	13.72%	15.77%	15.33%	13.77%
DP-3680	17.70%	21.14%	23.36%	17.36%

The results of testing the volume resistivity, outgassing, and glass transition temperature are shown in the Table 6.16.

TABLE 6.16 Properties of Flexipoxy Adhesives For Electrical Improvement Series

Candidate	Volume resistivity at 75°F, ohm-cm	Volume resistivity at 200°F, ohm-cm	Outgassing TML, %	Outgassing CVCM, %	Glass transition temperature
Flexipoxy 225.9, unfilled version	2.0×10^{14}	1.5×10^{11}	1.05	0.26	−60°C
Flexipoxy 225.9, filled version	3.2×10^{14}	3.0×10^{11}	0.45	0.13	−61.2°C
Flexipoxy 207, unfilled version	6.7×10^{13}	1.1×10^{11}	0.81	0.17	−28.5°C
Flexipoxy 207, filled version	No test	No test	(0.24–0.42)	(0.051–0.085)	No test
Flexipoxy 208, unfilled Version	8.1×10^{13}	1.3×10^{11}	0.75	0.11	−26.7°C
Flexipoxy 208, filled version	No test	No test	(0.22–0.38)	(0.033–0.055)	No test
Flexipoxy 1100, unfilled version	1.3×10^{14}	1.3×10^{11}	No data	No data	−60°C
Proseal 727	$1–3 \times 10^{12}$	$1–3 \times 10^{11}$	23.2	0.49	−40°C

Note: Values in parentheses are calculated from unfilled data.

The volume resistivities of the candidates are very similar, so nothing new was learned there. The outgassing properties of Flexipoxy 207 are good and of Flexipoxy 208 are excellent. Once made into a filled adhesive, Flexipoxy 208 should easily pass. The other thing of note is the markedly lower glass transitional temperature for Flexipoxy 225.9. This is apparently due to the high content of NC-547 epoxy resin.

6.6   THE STEERING COMMITTEE MAKES A FORMULATION SELECTION

Bob Mitsuhashi was very encouraged after seeing the properties of Flexipoxy 207 and Flexipoxy 208. He instructed us to get actual outgassing data on the filled versions and focus on finalizing these compounds so one of them could be qualified. Table 6.17 shows the outgassing results, volume resistivity, and glass transition temperature for actual filled and plasticized versions of Flexipoxy 207 and Flexipoxy 208:

TABLE 6.17   Properties of Flexipoxy 207 BF and Flexipoxy 208BF Paste Adhesives

Property	Flexipoxy 207 BF (filled)	Flexipoxy 208 BF (filled)
Outgassing: TML-WVR	0.34	0.27
Outgassing: CVCM	0.10	0.05
Volume resistivity at 75°F, ohm-cm	2.0×10^{14}	1.5×10^{11}
Volume resistivity at 200°F, ohm-cm	3.6×10^{14}	1.3×10^{11}
Glass transition temperature, °C	−28°C	−25°C

The two adhesives had very similar properties, but it was easy to determine which one to take into qualification testing. Passing the outgassing test is mandatory and Flexipoxy 207 BF was marginally failing the CVCM requirement. Flexipoxy 208BF was selected as the final candidate and efforts to qualify it begin.

6.7   THE FINAL CANDIDATE

Finally, we had combined the properties of non-outgassing and superior dielectric properties in one formulation. With volume resistivity greater than 1×10^{11} ohm-cm at 200°F, the problem of nano-amp leakage with MUX circuits disappeared. With the most important properties finally attained in a single formulation, it was time to advance to the qualification stage, where a full characterization would be undertaken, a production evaluation made, specifications written, and a decision to implement it would be made. The following is a description of the final candidate.

6.7.1 FLEXIPOXY 208BF FORMULATION

Table 6.18 shows the formulation for Flexipoxy 208BF paste adhesive.

TABLE 6.18   Formulation of Flexipoxy 208BF Thermal Transfer Adhesive

Ingredient	Type	Weight percent
Cardolite NC 547	Flexible epoxy resin	9.00
Cardolite NC 514	Flexible epoxy resin	9.00
Cardolite NC 513	Flexible epoxy diluent	0.95
PolyBd R45HT	Polybutadiene diol	3.80
325 mesh alumina	Conductive filler	66.00
Witco DP-3680	Flexible curative	7.25
Cab-O-Sil	Thixothrope	4.00

6.7.2 COMPOUNDING PROCESS FOR FLEXIPOXY 208BF PASTE ADHESIVE

6.7.2.1 PROCESS FOR A ONE KILOGRAM BATCH

Step 1: Dry 1 kg of 325 mesh alumina overnight at 300°F. Cool to ambient in a vacuum dessicator.

Step 2: Weigh into a clean mixing container:

(1) 90 g Cardolite NC 547
(2) 90 g Cardolite NC 514
(3) 9.5 g Cardolite NC 513
(4) 38 g Atochem PolyBd R45HT
(5) 72.5 g Witco DP-3680

Step 3: Thoroughly mix these ingredients, then add 660 g dried alumina from Step 1.

Step 4: Disperse alumina by hand mixing, then add 40 g Cabot Cab-O-Sil.

Step 5: Thoroughly mix by hand and transfer entire contents to the Ross planetary vacuum mixer.

Step 6: Mix under at least 300 milli-torr vacuum for 30 min; use cooling jackets needed to keep temperature at about 75°F.

Step 7: Dispense mixed paste immediately into 6 ounce Semco cartridges and flash freeze in liquid nitrogen. Store at −40°F until needed.

6.8   THE QUALIFICATION PROCESS

There had been some organizational changes that are worth discussing since the qualifications of Uralane 7760 and Flexipoxy 100 adhesives. Hughes Space and Communications Group (SCG) had previously relied heavily on the Technology Support Division (TSD) to do much of their materials and process engineering. However, they had made giant strides in becoming fully independent of us. David Chow, my former TSD Department Manager, had transferred to SCG in order to head this effort. Many of my M&P colleagues from TSD had followed him over SCG. One of the changes was to now create their own SCG specifications. The Hughes material specification (HMS) was superseded at SCG with their SCGMS designation. The Hughes Process Specification (HPS) became an SCGPS.

So, concurrent with the qualification testing, four new SCG specifications were being prepared:

(1) SCGMS 51088, material specification for the filleting adhesive
(2) SCGMS 51092, Material Specification for the Film Adhesive
(3) SCGPS 22104, process specification for flatpack bonding
(4) SCGPS 26037, process specification for fillet bonding

A division of assignments was made by the steering committee and specific projects of the qualification effort were assigned to TSD, SCG manufacturing, and SCG materials engineers.

6.8.1 CHARACTERIZATION OF THE FIRST BATCH

Characterization of the first batch of Flexipoxy 208BF adhesive was completed and reported on during the steering committee meeting on May 31, 1991. The following table compares the properties of Flexipoxy 208BF filleting adhesive with Uralane 7760 filleting adhesive and with Flexipoxy 100 if it were a filleting adhesive:

TABLE 6.19 Comparison of Filleting Adhesives versus Specification Requirements

Property	HMS 2308 requirement	Flexipoxy 100	Uralane 7760	Flexipoxy 208 BF
Work-life				
a) Hours extrudable	>2	>4	>2	>5
b) Lap shear strength	100–700 psi	300–400 psi	210 psi	391 psi
Storage life at −40°F	>3 months	8–16 months	>3 months	TBD
Color	Off-white	Off-white	Off-white	Tan
Volume resistivity,				
At 75°F, ohm-cm	$>1 \times 10^{11}$	$1\text{-}3 \times 10^{11}$	1.5×10^{15}	2.4×10^{14}
At 200°F, ohm-cm	None	About 10^9	$3\text{-}8 \times 10^{12}$	2.8×10^{11}
Dielectric strength, V/mil	>350	658	630	716
Dielectric constant	For info only			
1 KHz		10.7	5.10	5.56
10 KHz		9.7	5.00	5.34
100 KHz		8.3	4.80	5.16
1000 KHz		7.1	4.60	5.00
Dissipation factor	For info only			
10 KHz		0.081	0.018	0.011
100 KHz		0.095	0.020	0.012
1000 KHz		0.083	0.020	0.012
Outgassing				
TML-WVR, %	1.0 max	0.60	0.30	0.32
CVCM, %	0.1 max	0.04	0.00	0.05
Thermal conductivity,				
BTU/h-ft-°F	>0.25	0.33	0.43	>0.4*
Lap shear strength, psi	100-700	450	200	370

TABLE 6.19   *(Continued)*

Property	HMS 2308 requirement	Flexipoxy 100	Uralane 7760	Flexipoxy 208 BF
Glass transition temperature, °C	<4	−8	−63	−38
CTE > Tg, ppm/°C	<171	150	88	129
CTE< Tg, ppm/°C	<52	40	29	35
Durometer, Shore D	38 -58	48	88 Shore A**	44
Specific gravity	2.0-2.2	2.1	2.1	2.1
Solvent resistance				
a) % weight gain	<7.0	4.3	11.2	4.4
b) % thickness gain	For Info only	3.4	6.5	3.4

*Estimate based on alumina filler content.

**90 Shore A is usually equivalent to 45 Shore D.

Notice how similar are the electrical properties of Flexipoxy 208BF filleting adhesive and Uralane 7760 filleting adhesive, and how different are the electrical properties of Flexipoxy 208BF filleting adhesive and Hacthane 100 adhesive.

Table 6.20 compares the properties of Flexipoxy 208BF film adhesive with current film adhesive, Flexipoxy 100, and with former film adhesive, polysulfide adhesive.

TABLE 6.20   Comparison of Film Adhesives versus Specification Requirements

Property	HMS 2306 requirement	Flexipoxy 100 film adhesive	Polysulfide film adhesive	Flexipoxy 208BF film adhesive
Work-life				
a) Hours extrudable	>2	4	3	>2
b) Lap shear strength	100–700 psi	300–400 psi	210 psi	391 psi
Storage life at −40°F	>3 months	8–16 months	>3 months	TBD
Color	Off-white	Off-white	Off-white	Tan
Volume resistivity,				
At 75°F, ohm-cm	$> 1 \times 10^{11}$	$1–3 \times 10^{11}$	2.5×10^{12}	2.4×10^{14}
At 200°F, ohm-cm	None	2.5×10^{9}	3.5×10^{11}	2.8×10^{11}
Dielectric strength, V/mil	>350	625	484	896

TABLE 6.20 *(Continued)*

Property	HMS 2306 requirement	Flexipoxy 100 film adhesive	Polysulfide film adhesive	Flexipoxy 208BF film adhesive
Dielectric constant	<10.0			
1 KHz		11.3	8.3	4.8
10 KHz		9.8	8.1	4.7
100 KHz		8.4	8.0	4.5
1000 KHz		7.1	7.9	4.4
Dissipation Factor	<0.150	0.050–0.110		
10 KHz			0.0017	0.012
100 KHz			0.0024	0.012
1000 KHz			0.00930	0.013
Outgassing				
TML-WVR, %	1.0 max	0.60	23.1	0.40
CVCM, %	0.1 max	0.04	0.49	0.05
Thermal conductivity,				
BTU/h-ft-°F	>0.25	0.33	0.31	>0.4*
Lap shear strength, psi	100–700	400	290	340
Construction	Two step	Two step	One step	One step
Film thickness, mils	12–18	14–18	No data	14

Flexipoxy 208 BF film adhesive clearly has superior electrical properties than either of the other two film adhesives.

6.9 PROBLEMS ENCOUNTERED DURING QUALIFICATION

6.9.1 PROBLEM #1: BUBBLING OF THE FLEXIPOXY 208BF FILM ADHESIVE

A problem of bubbling was detected with the preparation of qualification samples during the curing cycle at 200°F. Our first task was to determine the source of the problem. FTIR analysis showed that the bubbles contained carbon dioxide gas. The source of the problem was the dry ice in the freezer, where the frozen premix is stored. Dry ice is actually the solid form of carbon dioxide. It tends to sublime at temperature above its freezing point, that is, it changes from solid to gas without a liquid intermediate state.

Dry ice is used in production to keep the film adhesive cold enough to remove the protective release films. We determined that the adhesive can be on dry ice for up to 40 min with no bubbling problem observed. We also learned that although Flexipoxy 208BF is more susceptible to bubbling than Flexipoxy 100 adhesive, both are susceptible to the problem.

The immediate solution is to enclose the film adhesive in vapor barrier bags. However, it was also determined that the choice of release film on the film adhesive is important to prevent bubbling. Polyethylene release film is far more effective in this regard than is Teflon-coated, non-porous (TFNP) film. The Steering Committee directed us to conduct a study of release films and select the best one. The criteria were (1) preventing bubbling due to dry ice, (2) ease of film adhesive fabrication, and (3) ease of release from adhesive. We in TSD M&P conducted the study.

The best release film was deemed to be 7 mil thick polyethylene film. It is vastly superior to TFNP film in preventing bubble formation.

6.9.2   PROBLEM #2: FAILURE OF FLEXIPOXY 208BF ADHESIVE TO PASS OUTGASSING

After all, the formulating effort to develop a non-outgassing thermally conductive adhesive with superior dielectric properties, it turned out they we were suddenly back at square one. Here is the story:

RESOLVING THE DIFFERENT CVCM RESULTS BETWEEN TSD AND SCG LABORATORIES

There developed a difference in the results of testing Flexipoxy 208BF for NASA outgassing between TSD M&P and SCG M&P laboratories. Although both labs found it to pass the TML-WVR requirement, SCG M&P found the CVCM values out of spec. They got a value of 0.12 %, which marginally fails the 0.10% maximum requirements. James Chow of TSD M&P conducted a round Robin test between the two labs to determine why the CVCM values are different. He learned that TSD uses semiclosed boat technique, whereas the open boat (SCG) is the correct technique. Consequently, Flexipoxy 208BF marginally fails the CVCM requirement of the NASA outgassing test. The following table shows the results:

TABLE 6.21 Flexipoxy 208BF Inter-Laboratory Round Robin Outgassing Test Results

Sample source	Tester	TML	WVR	TML-WVR	CVCM
Film, Lot #3	TSD	0.40	0.09	0.31	0.06
Film, Lot #3	SCG	0.44	0.10	0.34	0.13
Paste, Lot #2	TSD	0.41	0.08	0.33	0.08
Paste, Lot #2	SCG	0.46	0.08	0.38	0.12

REDUCING THE CVCM'S FROM FLEXIPOXY 208BF ADHESIVE

Both the TSD and the SCG laboratories agreed that Flexipoxy 208BF adhesive marginally fails the CVCM requirement of 0.10% maximum. The steering committee wanted the problem resolved by finding a way to pass the CVCM requirement without changing the formulation of Flexipoxy 208BF. The project was assigned to TSD M&P.

In order to determine which ingredients were responsible for the problem, we ran a simple experiment: A sample of NC-547 cured with DP-3680 was subjected to the standard outgassing test. The CVCM was 0.154%, a value exceeding the allowable 0.1% maximum. Our analytical chemist,Phil Magallanes, subjected the CVCM to FTIR Infrared analysis and compared the spectra to that of the NC-547 resin and to that of the DP-3680 curing agent. The CVCM spectra had the same spectra as the NC-547, but showed there was zero contribution from the DP-3680. Phil also determined that the NC-547 resin is composed of multiple components, each differing from one another in hydroxyl and epoxide content. The CVCM was very high in hydroxyl content.

Armed with this knowledge, we requested help from the manufacturer of the Cardolite epoxy resins, the Cardolite Company. As a result, Chris Ford, VP/ Technical Director of Cardolite, visited Hughes and was presented with our findings. He believed that the culprit is Cardanol. Special batches of Cardolite NC-547 were prepared for our evaluation without the diluent, NC-513.

Steve Lau used the cleaner Cardolite NC-547 resin to prepare batches of Flexipoxy 208BF adhesive. The existing formulation also specified 1% Cardolite NC-513 resin. Two additional batches were prepared, one batch reducing the NC-513 to 0.5% and the other batch to 0% NC-513. Table 6.22 shows the test results obtained from these three batches.

TABLE 6.22   Test Results on Modified Flexipoxy 208BF Adhesive

NC-513 content	CVCM, %	Volume resistivity at 75°F, ohm-cm	Volume resistivity at 200°F, ohm-cm	Lap shear strength, psi
1.0%	0.09	4.1×10^{14}	2.3×10^{12}	415 (SD = 32.9)
0.5%	0.084	1.3×10^{15}	1.1×10^{13}	447 (SD = 27.3)
0.0%	0.06	5.9×10^{14}	5.9×10^{12}	485 (SD = 70.0)

The steering committee selected the formulation with zero Cardolite NC-513 resin and ran a few tests to verify that it processed as well as the unmodified formulation.

6.10   COMPLETING THE QUALIFICATION OF FLEXIPOXY 208BF ADHESIVE

6.10.1   BATCH-TO-BATCH CONSISTENCY

The two additional batches were run through the same tests as was run on the first batch of Flexipoxy 208BF filleting and film adhesives and the results of all three batches were very consistent.

6.10.2   RADIATION TESTING

Dr. Tom Sutherland, SCG M&P, took care of irradiating volume resistivity specimens at one mega-Rad of gamma radiation. Both irradiated and control specimens were tested for volume resistivity from room temperature to −200°F and back to room temperature. Table 6.23 shows the test results at room temperature and 200°F for both control and irradiated specimens.

TABLE 6.23 Effect of Irradiation on Volume Resistivity of Flexipoxy 208BF Adhesive

Condition	Initial 75°F volume resistivity	200°F volume resistivity	Final 75°F volume resistivity
Control	2.8×10^{15} ohm-cm	4.6×10^{12} ohm-cm	$> 6 \times 10^{15}$ ohm-cm
Irradiated	6.3×10^{14} ohm-cm	3.3×10^{12} ohm-cm	$> 2 \times 10^{15}$ ohm-cm
Requirement	$>1 \times 10^{14}$ ohm-cm	1×10^{11} ohm-cm	$>1 \times 10^{14}$ ohm-cm

As is typical with this type of test, the test values tend to improve from the specimen's exposure to elevated temperatures due to additional cross-linking from the heat exposure. Also, notice that the irradiated samples had lower volume resistivities than did the control sample. Even so, those test values were well above the requirements. The conclusion is that there is little to nil radiation damage to Flexipoxy 208BF, which was detectable by this test.

6.10.3 FLEXIPOXY 208BF ADHESIVE COMPATIBILITY WITH SOLDER FLUX

A compatibility study of Kester 1350 HA solder flux on both the paste and film adhesive forms of Flexipoxy 208BF was conducted by our TSD M&P laboratory. FTIR spectroscopy was used to examine the flux alone, uncontaminated Flexipoxy 208 BF paste, and film adhesive and flux contaminated paste and film adhesives. The FTIR spectra were run on all of these specimens in both the uncured and the cured state. All of the specimens cured to a strong leathery state. The adhesive was deemed compatible with solder flux.

6.10.4 GLASS-BODIED DIODE TESTING

Lydia Simanyi of SCG M&P conducted the evaluation of Flexipoxy 208BF filleting adhesive on glass-bodied diodes and determined that it passed.

6.10.5 COMPATIBILITY TESTING

Jim Cammarata of SCG M&P found that Flexipoxy compatible with toluene is acceptable. He also found that bonding of the adhesive to glass fiber reinforced polyimide composite or diallyl phthalate (DAP) was acceptable.

6.10.6   OTHER QUALIFICATION TESTS

Many other tasks were conducted in order to complete the qualification process.

1. Freezer storage shelf life was determined.
2. The time to cure the adhesive was determined at temperatures from room temperature up to 200°F.
3. A new and better sag test for the filleting adhesive was developed and added to the material specification.
4. Coefficients of thermal expansion were developed above and below the glass transition temperature using the quartz dilatometer. The data were obtained for the stress engineers to run a finite element analysis of the new adhesive in a bonded flatpack configuration. However the stress analysis was deemed unnecessary based on the superior properties of the Flexipoxy 208BF adhesive when compared with Flexipoxy 100 properties.
5. FTIR spectra were developed on each of the liquid ingredients. Refractive index, which is a very precise measurement of purity, was run on all of the transparent liquid ingredients.

6.11   PAPER AND PRESENTATION

I prepared a paper and presentation for the 6th Annual International SAMPE Electronics Materials and Processes Conference held June 23–25, 1992. The paper was titled, "The Evolution of Space Qualified Thermal Transfer Adhesives." The authors were Ralph D. Hermansen, Steven E. Lau, and Robert B. Mitsuhasihi. The paper discussed how thermal transfer adhesives for space electronics evolved from polysulfides with their serious outgassing problems to Uralane 7760 and Hacthane 100 for filleting discrete components and Flexipoxy 100 film adhesive for flat-bodied components. Although these new adhesives were a major improvement, further improvements were needed. The result of very innovative formulation development work was Flexipoxy 208BF, which resolved the outstanding problems. I presented the paper in Baltimore.

6.12 PATENT FOR A SUPERIOR THERMAL TRANSFER ADHESIVE

Steve Lau and I applied for a patent on our superior thermal transfer adhesive on January 6, 1994. U.S. 5.367,006 was awarded on November 22, 1994. In the abstract of the patent, the novelty of the invention is described as attaining a combination of unique properties, namely, volume resistivity of 10^{14} ohm-cm, flexibility to $-55°C$, maximum thermal conductivity, excellent adhesion, and reworkability.

6.13 IMPLEMENTATION

The manufacturer of Cardolite NC-547, Cardolite Corporation, decided it did not wish to manufacture the cleaned up version of NC-547. Our organic chemistry group had attempted to remove volatiles from the NC-547 resin to no avail. The Flexipoxy 208BF adhesive project was put on hold at that point.

To say I was disappointed with the curtailment of the project is a huge understatement. Moreover, I was frustrated because I could not think of a way to save the huge investment SCG had made in developing this new adhesive. Looking to the positive side, Steve Lau and I had made huge gains in our ability to formulate flexible epoxies for electronic assembly and many other uses.

KEYWORDS

- **Flexipoxy**
- **cardolite epoxy resins**
- **volatile condensable matter**

PART III
Other Custom-Formulated Compounds for Aerospace Electronics Applications

CHAPTER 7

RADIOPAQUE ADHESIVE/SEALANT

CONTENTS

This section contains five chapters, each of which describes an engineering problem, where custom formulating offered a solution to the problem. In Chapter 7,
development of a novel radiopaque adhesive allowed voids to be seen in an adhe-
sive/sealant sandwiched between aluminum parts. The application was a nutation
damper on the GOES satellite. In Chapter 8, an epoxy impregnant for satellite
high-voltage power supplies needed to be reformulated to cure at a lower temperature to reduce residual stresses from developing. In Chapter 9, a special adhesive
was needed to bond cyanate ester composite parts to one another in a microwave
channeling device. The adhesive had to be platable using the same etching processing as for the composite parts. In Chapter 10, very high thermal conductivity
is needed in an adhesive for a naval weapon system. The adhesive must also be of
controlled strength for mission survival yet easy removal if necessary. In Chapter
11, flexible epoxy adhesives for three different functions are formulated to have
room temperature storage stability as one-component systems. All of these inventions were patented.

7.1   THE PROBLEM

7.1.1   THE GOES NUTATION DAMPER

In March 1986, the Space and Communications Group of Hughes Aircraft
had a problem with their nutation damper on the GOES satellite program.
The nutation damper consisted of an aluminum tube, which was bent to
form a complete circular ring. A 2-inch-long aluminum sleeve fits over the
area where the ends of the tube join together. The aluminum is 7075 T73.
An epoxy adhesive, Epiphen 825A, was used to bond the aluminum parts
together and simultaneously seal the joints to provide an air-tight seal. The
tube will contain a gas.

The problem was that too often the epoxy adhesive did not provide a
gas-tight seal. The leaks were due to entrapped bubbles, knit-lines, and
voids in the adhesive. More troubling is the fact that there was uncertainty
about any of the nutation dampers. The GOES satellite sees 300 thermal
cycles on orbit from a temperature as low as −63°C to as high as 30°C.
These temperature changes could generate a leak due to the stresses generated as parts expand and contract. In addition to the current part, there has
been an ongoing problem with the nutation damper. Of the previous parts,
there was one good one, one 70% filled, one 60–70% filled, and one 80%
filled.

7.1.2 THE ORIGINAL IDEA FOR A RADIOPAQUE ADHESIVE

Roamar Predmore of NASA asked Dr. Tom Sutherland of Hughes Aircraft if it might be possible to develop an adhesive which would be more radiodense than the aluminum. He believed that if the adhesive could be seen in the nutation damper by means of x-ray inspection, voids could be detected. Moreover, there is no easy non-destructive test to ascertain the quality of the adhesive seal. X-ray inspection would be an excellent test to spot voids except that one cannot see the adhesive inside the aluminum with x-rays because aluminum is more radiopaque than the adhesive. The question posed to me was "Is it possible to modify the Epiphen 825A adhesive so that it is more radiodense than the aluminum tube and sleeve, while preserving the desirable adhesive and sealant properties?"

7.2 THE GOAL

The goal is to formulate a radiopaque adhesive based upon Epiphen 825A adhesive, which would still retain its most important adhesive properties including good processing characteristics, high lap shear strength, and non-outgassing properties.

7.3 MY PREVIOUS APPLICABLE EXPERIENCE

7.3.1 MY X-RAY EXPERIENCE

Although hardly an expert on x-ray inspection, my background did include engineering experience on products where radiation is a factor. My first three jobs were with prime contractors to the Atomic Energy Commission (AEC). Those jobs were with Argonne National Laboratory, ACF Industries, Albuquerque Division and Sandia National Laboratory. At Sandia, I used a test called DXT to measure the density of high-density plastic foams using x-ray inspection. Essentially, the density of the foam was determined by comparing the x-ray absorption of a foam sample with a standard included in the test. I was also fortunate to take a course in radiochemistry during my graduate work at the University of New Mexico. However, that course focused mainly on nuclear reactions.

7.3.2   EXPERIENCE WITH EPIPHEN 825A ADHESIVE

Moreover, I was very familiar with the adhesive being used, namely, Epiphen 825A adhesive. This room temperature curing structural adhesive has been the workhorse for joining parts in the Space and Communications Group. During my first year in the Space and Communication Group of Hughes Aircraft in 1980, I was heavily involved in an intensive effort to improve the reliability of adhesive bonding of satellite parts. This effort caused me to become intimate with the adhesive. Then 2–3 years later, I was deeply involved in investigating a quality issue with the adhesive's supplier and a disturbing trend toward lower lap shear strengths. An extensive study of the adhesive components was undertaken in which I played a key role.

7.3.3   MORE ABOUT EPIPHEN 825A ADHESIVE

Epiphen 825A adhesive is the adhesive used to bond and seal the nutation damper ring. The adhesive is qualified to Hughes Material Specification 20-1805. The original adhesive, which later became Epiphen 825A, was developed under Air Force Contract in the early 1960s. Borden Chemical picked up the formulation and modified it to Epiphen 825A. Hughes Aircraft Company subsequently qualified it for general use. The formulation was sold to Haven Chemical when Borden Chemical went out of the business, but eventually Haven Chemical also went out of business. Finally, we qualified a third supplier in 1982, namely, Monomer-Polymer and Dajac Company of Trevose, Pennsylvania.

The adhesive is sold as a four-component system. Epiphen 825A is very commonly used by HAC for room temperature structural bonding. There is an arrangement with Ablestik Adhesive Company of Gardena, California to combine the four components to make a one-component adhesive and to provide it to HAC as a frozen premix adhesive. Working with a one-component adhesive eliminates the need for HAC technicians to weigh, mix, and degas the adhesive each time they need to use it. Even more significant is the assurance that any frozen premixed Epiphen 825A has been tested and approved prior to its use on flight hardware.

7.4   TECHNICAL APPROACH

Radiopacity is directly related to the effective atomic number of the substance. Technologists often refer to the atomic number as Z of the substance, derived from the German word "zahl" which means number. We know that the adhesive cannot be seen with x-ray inspection because it is less radiodense than aluminum. Aluminum has an atomic number of 13. The adhesive is predominately organic compounds comprising carbon, oxygen, nitrogen, and hydrogen. Their atomic numbers are 6, 8, 7, and 1, respectively. The polysulfide modifier contains sulfur with an atomic number of 16. The filler is mainly silicon (14), oxygen (8), aluminum (13), and potassium (19). Table 7.1 below shows the breakdown of the four-component adhesive by the chemistry of the ingredients and their atomic contributions. Note that the Mica filler is about 1/4 of the formula by weight but much less than that by volume due to its higher density. Radiopacity is dependent on the volume effect, not on weight.

TABLE 7.1   Description of Epiphen 825A Adhesive

Adhesive component	Parts by weight	Specific gravity	Ingredient description	Elements in the ingredient
Epoxy resin	25	1.22	Epoxy novolac resin + low viscosity tri-epoxide resin	H, C, O
Polysulfide modifier	3	1.27	Mercaptan-terminated polyether	H, C, O, S
Mineral filler	10	2.6-2.75	325 mesh Talc powder	Si: 47.9% as SiO_2
				Al: 33% as Al_2O_3
				K: 10% as K_2O
Polyamine converter	4	0.90	Mixture of two aliphatic amines	H, C, N, O

7.5   THE STEPS IN DEVELOPING A RADIOPAQUE ADHESIVE

7.5.1   LOCATING A HIGH Z POWDER

I visited TSD's chemical storage area and looked for a powder containing the highest Z number that I could find. Having a high Z number (atomic number of the element) is the key to attaining high radio-opacity. I found a Tungsten salt in powder form. Tungsten has Z of 74. It should be noted that Tungsten also has a specific gravity of 19.35. Thus, much less volume of Tungsten is attained per unit weight as we replace the Epiphen 825A filler with Tungsten or its salt.

7.5.2 TOM SUTHERLAND ANALYZES THE PROBLEM MATHEMATICALLY

Dr. Tom Sutherland was both an academic and an applied physicist. Starting with the knowledge of the GOES nutation damper configuration, the composition of Epiphen 825A adhesive, and the fact that we had Tungsten powder at our disposal, he set out to determine mathematically what level of Tungsten powder would be necessary to modify the adhesive adequately to attain a radiopaque property.

In the nutation damper, the adhesive is sandwiched between two layers of aluminum, L1 and L2, where L1 is 0.063 inch thick aluminum and L2 is 0.016 inch thick aluminum. The adhesive is nominally 0.005 inch thick. He assumed x-ray wavelengths of either 0.15 or 0.5 Angstrom units. I am not going to repeat the pages of differential equations that Tom used to prove the feasibility of our "Tungsten as a filler in Epiphen825A" concept, but if the reader would like to try it at home, Tom started with Beer's law.

7.6   EXPERIMENTAL

7.6.1   EXPERIMENT PLAN

In the aerospace industry, theoretical calculations are important, but the responsible design engineers want to see test data that confirm the theory and so, Epiphen 825A adhesive was made in several different versions, where the Tungsten powder replaced increasing percentages of the mica filler. Table 7.2 shows the trials expressed in volume units, because the attenuation is a volume controlled phenomena.

TABLE 7.2   Composition of Modified Epiphen 825A Test Adhesives

Specimen I.D.	Volume ratio of polymer/ filler	Volume % tungsten powder in total filler	Volume % mica powder in total filler
5-95	67/33	5	95
10-90	67/33	10	90
25-75	67/33	25	75
50-50	67/33	50	50
75-25	67/33	75	25

Samples of bonded aluminum were prepared with the test adhesives and X-ray photos were taken to see at which level the adhesive could be seen through the aluminum. Two inch × two inch aluminum sandwich coupons were prepared as follows:

TABLE 7.3   Configuration of Test Specimens

Layer numbers (from top to bottom)	Composition of layer
1	0.063-inch-thick aluminum
2	0.014-inch-thick test adhesive
3	0.016-inch-thick aluminum

The inspection was conducted with a Hewlett-Packard Model 4384N X-ray machine. Two different x-ray wavelengths were used in the testing, namely, 0.15 and 0.5 angstroms.

7.6.2   TEST RESULTS

The test results show that the adhesive is visible through aluminum at 15% volume replacement of the filler at 0.5 angstroms. However, at least 40% volume replacement was needed for 0.15 angstrom x-rays. As expected, the higher the percentage of Tungsten in the filler, the more clearly the voids could be identified.

A GOES nutation damper was also injection bonded with our radiopaque adhesive. The voids in the adhesive were clearly visible.

To assure that the filler modifications did not reduce adhesive strength, the test adhesives were tested for lap shear strength per ASTM D1002. All of the candidates exceeded the requirement of 2500 psi.

7.7   CONCLUSION

The concept of formulating a radiopaque adhesive, which allows the non-destructive inspection of bonded joints within aluminum parts, was both theoretically and experimentally proven to work. The good processing characteristics of the radiopaque adhesive were maintained, and the lap shear strength of the radiopaque was equivalent to the original adhesive. Both values easily passed the specification requirement.

7.8   THE PATENT

The U.S. patent 4,940,633 entitled "Method of Bonding Metals with a Radio-Opaque Adhesive/Sealant for Void Detection and Product Made" was awarded to Ralph D. Hermansen, Thomas H. Sutherland, and Roamer Predmore on July 10, 1990. The filing date for the patent application was May 26, 1989. The patent abstract reads as follows:

A method and structure for providing radio-opaque polymer compounds for use in metal bonding and sealing. A powder filler comprising a high atomic number metal or compound thereof is incorporated into a polymer compound to render it more radio-opaque than the surrounding metal structures. Voids or other discontinuities in the radiopaque polymer compound can then be detected by x-ray inspection or other non-destructive radiographic procedure.

KEYWORDS

- **GOES satellite**
- **Epiphen 825A**
- **radiopaque adhesive**

CHAPTER 8

LOW EXOTHERM, LOW-TEMPERATURE CURING, EPOXY IMPREGNANTS

CONTENTS

8.1   HIGH-VOLTAGE ENCAPSULATION

Except for this single patent, you would never know that I worked on high-voltage encapsulant problems. Yet they may be as much as 10% of my time at Hughes Aircraft Company after I transferred to the Technology Support Division. I could not work on electrical problems during the first 5 years because my area was structural assembly and the electrical assembly problems belong to someone else.

I loved working on high-voltage encapsulant problems because they were so challenging and I was learning about an entirely new area of technology. Most of these assignments came from the Space and Communications Group, although some came from the Missile Systems Group. The majority of these problems involved high-voltage power supplies.

There is a world of difference between high-voltage encapsulation and low-voltage encapsulation. High voltages tend to destroy dielectric materials if they can find a weakness. The weakness may be an entrapped bubble, a knit line, a delamination, or a host of other defects. Once arcing can begin, the resulting carbon path is conductive and the path continues growing until it finds a ground.

8.2   ENCAPSULATION USING IMPREGNANTS

8.2.1   STIRRED-IN FILLERS OR FIBERS

Where good thermal conduction or high mechanical strength is required in the encapsulant, there are ways to achieve these properties by making the encapsulant a composite material. For thermal conduction, the encapsulant is often a composite of a dielectric, mineral filler and a polymeric matrix (usually an epoxy). For strength, the use of glass cloth or fiber and a polymeric matrix (usually epoxy) is a suitable solution.

Although conductive filler or short glass fibers could be stirred into a liquid epoxy system, the resultant composite tends to have entrapped bubbles due to its high viscosity. Moreover, the amount of filler possible in the composite is quite limited. Particularly, there is a limit to how much short fiber for reinforcement can be added to a liquid polymer before flow properties of the filled encapsulant are lost. However, there is also a limit for mineral fillers such as silica, glass beads, calcium carbonate, clay, talc, and alumina particles. The viscosity of the mixture rises rapidly with increasing filler content making the entrapment of voids, bubbles, and porous knit lines unavoidable. Such inclusions are weaknesses in the dielectric encapsulant and sites where breakdown under high voltage can begin.

8.2.2 IMPREGNATING FILLERS AND FIBER REINFORCEMENTS

However, an alternate approach is to put the fibers or fillers into the cavity for the encapsulant and introduce the liquid polymer into the interstices of the fiber or filler via vacuum impregnation. Air is removed from the cavity and the liquid polymer is driven into the matrix due to a combination of atmospheric pressure and capillary action. The impregnant must be of low viscosity for this technique to work and the work-life of the resin/hardener mixture must be long enough for the impregnation to take place before the viscosity rises too much.

One example of this process common to high-voltage encapsulation is the wrapping of components with glass cloth, followed by vacuum impregnation with an epoxy resin/hardener mixture. When the impregnant is cured, the resultant composite has the high mechanical properties of a reinforced plastic.

Another example of the process common to high-voltage encapsulation is to fill the cavity with alumina particles vibrating the setup to achieve maximum packing density. Next, the alumina is vacuum impregnated with an epoxy resin/hardener mixture. When the impregnant is cured, the resultant composite has a higher coefficient of thermal conductivity, lower coefficient of thermal expansion, and less shrinkage than an unfilled encapsulant.

8.2.3 STRESS ANALYSTS ARE PART OF THE PROBLEM-SOLVING TEAM

On most of the high-voltage encapsulant problems, I found myself working closely with a stress analyst. The function of the stress analyst is to build a mathematical model of the device in question, input the needed material properties of the different materials, and run computer simulations of the device to see how it responds. Usually, there is a lot of materials testing involved in obtaining the data for the computer model. Typical properties include tensile strength, modulus, and elongation as well as compressive and shear properties. These data may need to be obtained at different temperatures. Also, thermal conductivity and thermal expansion coefficients are needed over a defined temperature range. Poisson's ratio is a property which tells how the material deforms under applied stress. It is sometimes needed too.

8.3   THE SPECIFIC HIGH-VOLTAGE PROBLEM

For the SAP 391 Electronic Power Conditioner (EPC), the current epoxy impregnant must be cured at temperatures of 160°F or higher to achieve an adequate cure. However, residual stresses arise at this cure temperature which can contribute to high-voltage breakdown of the dielectric. David Sandkulla was a stress analyst assigned to the SAP 391 high-voltage breakdown problem. He had run a two dimensional stress analysis and determined that due to the 160°F cure of the impregnant, residual stresses in the encapsulant were excessive when the satellite saw temperatures dipping down to 0°F. He had also calculated that a cure temperature of 130°F or lower would adequately lower these stresses.

8.4   THE TECHNICAL APPROACH TO SOLVING THE PROBLEM

8.4.1   THE CURRENT IMPREGNANT

The currently used impregnant is a mixture of Epon 815 and menthane diamine. We can refer to it as 815/MD. This would have been the perfect impregnant for high-voltage applications, if only it cured at 130°F instead of the 160°F, which was actually the temperature required to properly cure it. In order to create a 130°F curing impregnant, our technical approach was to look for alternate diamine curatives. Alternative curatives might allow lower cure temperatures without sacrificing other needed properties.

8.4.2 STRAIGHT CHAIN, CYCLO-ALIPHATIC, OR AROMATIC DIAMINES

Menthane diamine is a cyclo-aliphatic diamine, and it has proven to cure Epon 815 epoxy resin to a polymer with adequate thermal stability for the application. Table 8.1shows the three main categories of diamines. Straight chain diamines cure at lower temperatures, but lack the thermal stability of cyclo-aliphatic diamine- cured epoxies. Aromatic diamines have excellent thermal stabilities, but their minimum cure temperatures are far too high.

TABLE 8.1 Major Types of Diamine Curatives and Their Characteristics

Type of diamine	Cure temperature range	Thermal stability
Straight chain, aliphatic	75°F and higher	Inadequate
Cyclo-aliphatic	120–200°F	Adequate for the application
Aromatic	>300°F	Excellent

So we should focus on cyclo-aliphatic diamines. Now, amines link to epoxide groups via an active hydrogen atom attached to the nitrogen atom. Primary amines have two active hydrogens, whereas secondary amines have one active hydrogen. We are only interested in primary diamines, so our search is for cyclo-aliphatic primary diamines.

8.4.3 STERIC HINDRANCE IS A KEY FACTOR

What is steric hindrance? When bulky organic groups, such as methyl, ethyl, phenyl, cyclohexyl, and others, are adjacent to a reactive group on a molecule, they tend to reduce the probability of that group reacting within a given time. Specifically, bulky groups adjacent to an amine group reduce the reactivity of that amine group reacting with an epoxide group in a given time at a given temperature. If we raise the temperature, there is greater probability of a reaction in a given time.

Menthane diamine requires a 160°F or greater cure temperature because both of its primary amine groups are sterically hindered. One of the amine is shielded by a methyl group and a cyclohexane group and the other amine is shielded by two methyl groups. If we can find a cyclo-aliphatic diamine with less steric hindrance than menthane diamine, it will properly cure at a lower temperature.

8.4.4 IMPREGNANT WORK-LIFE AND PEAK EXOTHERM

However, if our new diamine causes too fast of a reaction, new problems develop. The viscosity of the impregnant will rise too rapidly and the impregnant may not completely impregnate the filler or cloth material before it hardens. Another concern is the exothermic heat developed from the amine/epoxide reaction. If the reaction occurs too quickly, the heat of the reaction could create a safety hazard. Like in the story of Goldilocks and the three bears, we need a curative that is not too hot, not too cold, but just right.

8.5 EXPERIMENTAL—EXOTHERM STUDIES

8.5.1 EPOXY AND DIAMINE REACTANTS

Eighteen epoxy formulations were evaluated for the viscosity of the mixed reactants and for peak exotherm temperature and the elapsed time to reach it. Table 8.2 shows the reactants which were used in the evaluation. PACM and MXDA are unhindered diamines. PACM is a cyclo-aliphatic diamine, whereas MXDA is an aromatic compound, which has the reactivity characteristics of a cyclo-aliphatic diamine.

TABLE 8.2   The Reactants and Their Properties

Reactant ID	Chemical name	Epoxide equivalent	Active H equivalent
Epon 828	Diglycidyl ether of bisphenol A	185-192	none
Epon 815	89% Epon 828 + 11% butyl glycidyl ether	180-195	none
Heloxy 67	Diglycidyl ether of 1,4-butane diol	124-127	none
Menthane diamine (MD)	4-(2-aminopropan-2-yl)-1 -methylcyclohexan-1-amine	none	42.5
PACM	para-bis (amino-cyclohexyl) methane	none	52.5
MXDA	Meta-xylylene diamine	none	34

Some of the ingredient chemical structures are shown below:

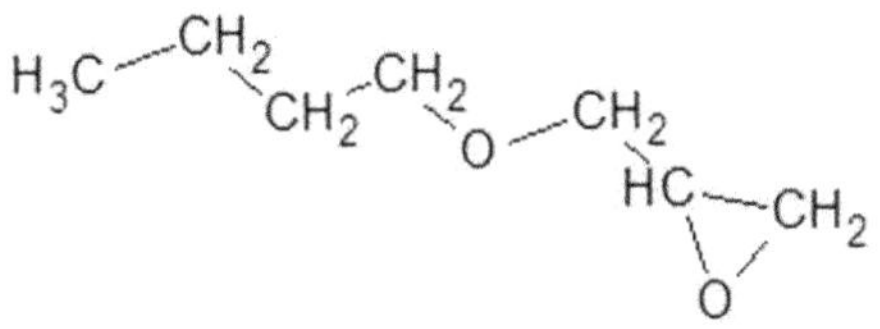

FIGURE 8.1 Structure of butyl glycidyl ether.

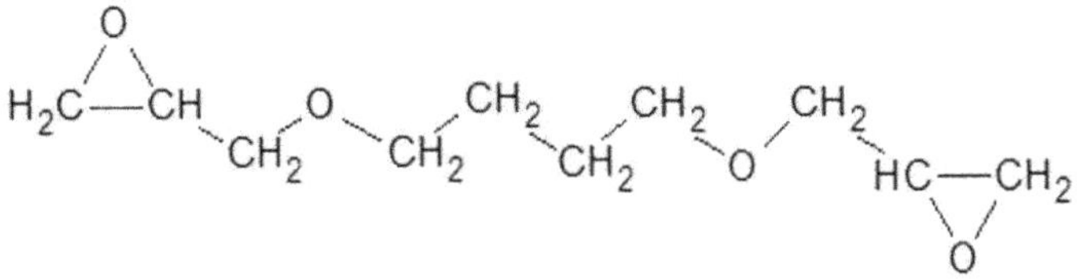

FIGURE 8.2 Structure of Heloxy 67 epoxy resin.

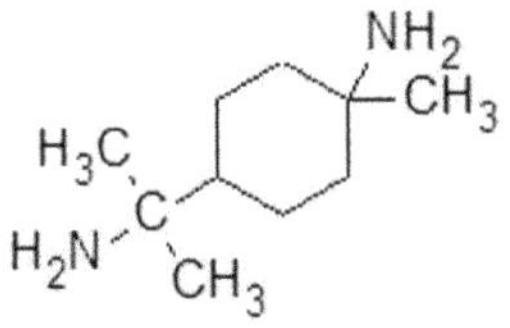

FIGURE 8.3 Structure of menthane diamine curative.

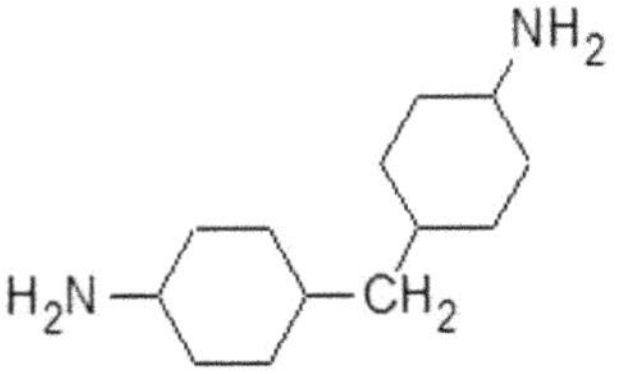

FIGURE 8.4 Structure of PACM curative.

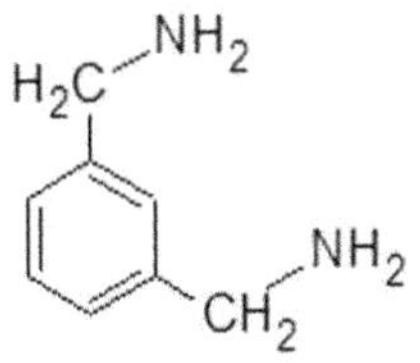

FIGURE 8.5 Structure of MXDA.

8.5.2 DESCRIPTION OF THE EXOTHERM TEST

Epoxy resin systems are known to be sensitive to the amount of material being mixed in terms of runaway exotherm. While one quantity of material may be safe, a greater quantity may be unsafe in that the heat of reaction may scorch or char the reacting polymer, burst from the container due to the expanding hot gases of decomposition, and actually propel hot pieces in the air. When offering a more reactive epoxy system to replace a less reactive one, the danger of runaway exotherm must be determined.

The reason that mass sensitive exotherm is a problem is that reactive polymers have extremely low coefficients of thermal conductivity. They are about one-tenth that of glass, which is quite low itself. Consider that a person can hold a glass rod in one's finger while its molten end a couple inches away is red hot and yet one feels no discomfort. The polymer is 10 times more insulative. Next, imagine perfectly spherical onions of different diameter. These onions represent reacting polymer of different masses. The larger the diameter, the greater the number of insulating layers surround the center. In the case of the onions, there are more layers on a bigger onion than on a smaller one. The heat of the reaction cannot escape, so as the temperature climbs, the reactions occur faster and faster.

In order to compare different formulations of epoxy impregnants for tendency to exotherm, a test was developed. One pint metal paint cans are readily available. So, the metal pint can become the standard container for the test. Sufficient epoxy resin and hardener of the test material were weighed out and mixed to provide 473 cc for the test. The test material was immediately transferred to the pint can and a thermocouple was securely positioned at the center of mass. A strip chart recorder was activated to record the temperatures provided from the thermocouple. When the test is complete, the peak exotherm temperature and the elapsed time to peak temperature are read off the chart.

8.5.3 RESULTS FROM THE EXOTHERM TEST

The tables below show the formulations of the trials and the results of testing mixed viscosity and exotherm behavior.

TABLE 8.3   Exotherm Studies Series #1

Ingredient or property	Trial 1	Trial 2	Trial 3	Trial 4	Trial 5
Epon 828, pbw	100	100	100	100	100
PACM, pbw	0	27.5	27.5	27.5	27.5
MD, pbw	22	0	0	0	0
Alumina, pbw	0	0	50	100	200
Viscosity, cps.	2450	2650	4600	6800	25,000
Peak exotherm temperature, °F	91	452	308	234	120
Time to peak exotherm, h	20	2.1	2.4	3.0	5.5

Comparing Trials 1 and 2, we see that there is no concern of runaway exotherm using menthane diamine curative but there is a serious problem using PACM. Trials 2 through 5 show that with increasing amount of alumina filler in the formulation, the exotherm temperature decreases and the time to peak exotherm increases.

Notice that when the impregnant is filling the airspace between filler particles or the air space in glass cloth, its work-life is longer and peak exotherm lower than the unfilled epoxy in the mixing container. The danger of an exotherm hazard is greatest in the mixing container.

TABLE 8.4   Exotherm Studies Series #2

Ingredient or property	Trial 6	Trial 7	Trial 8	Trial 9	Trial 10	Trial 11
Epon 815. pbw	100	100	100	100	100	100
PACM, pbw	0	7	10.5	14	18.2	28
MD, pbw	28	21	17.5	14	9.8	0
Alumina, pbw	0	0	0	0	0	0
Viscosity, cps	300	330	350	360	380	425
Peak exotherm temperature, °F	90	137	295	311	362	400
Time to peak exotherm, h	25	10.5	7.4	5.3	3.9	0.13

As we progress from Trial 6 through to Trial 11, the curative blend changes in steps from all MD to all PACM. All of the trials were unfilled. All of the trials had low enough of a viscosity to be an impregnant. The noteworthy thing is that the peak exotherm temperature changes in a predictable pattern as the ratio of the two curatives changes. The higher the percentage of PACM, the higher the peak exotherm temperature. Similarly, the time to peak exotherm is also predictable. The higher the percentage of PACM, the quicker the peak exotherm is observed.

TABLE 8.5 Exotherm Studies Series #3

Ingredient or property	Trial 12	Trial 13	Trial 14	Trial 15	Trial 16	Trial 17	Trial 18
Epon 828, pbw	100	100	100	100	100	85	85
Heloxy 67, pbw						15	15
PACM, pbw	15	0	0	0	0	0	0
MD, pbw	10	5.7	5.7	10	10	28	28
MXDA, pbw	0	13.3	13.3	10	10	0	0
Alumina, pbw	0	0	50	0	50	0	50
Viscosity, cps.	2350	1490	2090	1320	1960	1100	1600
Peak exotherm temperature, °F	134	425	301	390	260	90	85
Time to peak exotherm, h	6.3	3.3	2.8	3.2	3.7	21	19

Comments:

1. If we compare Trial # 12 with Trial #10 of the previous series, we can learn something about using Epon 828 in place of Epon 815. For one, the impregnant viscosity of Trial # 12 is much higher than that of Trial #10. Actually, Epon 815 is Epon 828 with a mono-epoxide diluent added to lower the viscosity. Yet, that small change made a big change in exotherm properties. The Epon 828 resin produced a much lower peak exotherm than the Epon 815 resin. Unfortunately, the Epon 828 formulation (Trial # 12) was too viscous to be an impregnant.

2. Comparing Trials # 13 and # 15 as part of an unfilled series, we see that the presence of MXDA in the formulation causes a high exotherm temperature. Trials # 14 and # 16 would be the same series except there is 50 pbw alumina filler in the formulations. The peak exotherm temperatures are lower due to the heat-absorbing inert filler particles. The times to peak exotherm are not lengthened as in unfilled Series #2. The alumina particles help the heat from the center of mass to be conducted away better than what is possible in an unfilled formulation, which is highly insulative.

3. If we compare Trial # 17 and Trial # 11 of the previous series, we are looking at Epon 828 diluted with a monoepoxide and cured with PACM (Trial #11) and Epon 828 diluted with a diepoxide diluent (Heloxy 67) and cured with PACM (Trial #17). Although Trial #17 has almost three times higher viscosity than Trial #11, its peak exotherm is vastly lower, and time to peak exotherm is vastly longer than those values for Trial #11.

8.6   EXPERIMENTAL—COMPLETENESS OF CURE STUDIES

Concurrent with the 'Eighteen Trials Experiment' discussed above, we evaluated Epon 815/PACM trials for tendency to cure at 100°F and 120°F, using Shore D durometer as an indicator of complete cure. The stoichiometric amount of PACM is 28.5 parts per 100 parts of Epon 815. We looked at how the cure would be affected if the mix ratio were either higher or lower. Table 8.6 shows the results:

TABLE 8.6   Durometer Climb for Epon 815/PACM Trials at Different Cure Temperatures

Formulation	Cure Temperature, °F	5 h	8 h	24 h
100 815/23.5 PACM	100	75 D	86 D	85 D
100 815/28.5 PACM	100	75 D	85 D	88 D
100 815/28.5 PACM	120	85 D	88 D	88 D
100 815/32.5 PACM	100	75 D	85 D	88 D
100 815/34.5 PACM	100	82 D	86 D	87 D

Comments:

1. A full cure appears to be attainment of 88 Shore D.
2. PACM used at 28.5–32.5 parts per 100 resin attained a cure within 24 h at 100°F.
3. PACM used at 28.5 parts per 100 resin attained a cure within 8 h at 120°F.
4. PACM used between 23.5 and 34.5 part per 100 Epon 815 resin cured to become a rigid epoxy plastic within 5 h at 100°F.

8.7   MY RECOMMENDED SOLUTION

For the SAP 391 Program, we now had a way to cure the impregnant at a temperature lower than 160°F and reduce the residual stress, which led to high-voltage breakdown of the units. The following table summarizes the results:

TABLE 8.7   Curative Blend versus Properties

PACM/menthane diamine ratio in curative blend	Minimum cure temperature, °F	Peak exotherm, °F	Time to peak exotherm, h
0:100	160	90	25 h
25:75	150	137	10.5 h
37.5:62.5	115–135	295	7.4 h
50:50	105–125	311	5.3 h
65:35	100–115	362	3.9 h
100:0	90	400	8 min

Note: This is the same data from Table 8.4, where the epoxy resin was Epon 815 and 28 parts of the curative blend was used with 100 parts of Epon 815 resin.

RECOMMENDATION

A lower curing impregnant could be selected from the data at around 50/50 curative blend composition. The cure temperature could be kept under 130°F and the exotherm of an unattended pint mixture would be far under the decomposition temperature of the material and it would take over 5 h to occur.

We had also recommended that SCG fund an effort to qualify the new impregnant by determining important properties such as mechanical properties, electrical properties, outgassing, and so on. Instead, they decided to attempt to solve the problem in a different manner. It can be an expensive process to re-qualify the hardware, and material changes can cause concern with the customer.

8.8   THE PATENT

U.S. Patent 5,350,779 entitled, "Low Exotherm, Low Temperature Curing, Epoxy Impregnants" was filed for on June 29, 1993 and was awarded to Ralph D. Hermansen and Steve Lau on September 27, 1994.

8.9 PRESENTATION

One area of investigation not mentioned in the chapter was a study of how fast impregnants of different viscosities penetrate fillers of different particle size and shape. These data were included in a talk I gave entitled, "The Role of Fillers in Encapsulants and Potting Compounds." The paper and presentation were given at the 22nd International SAMPE Technical Conference in Boston, November 6–8, 1990.

8.10 POST ANALYSIS

As I was reviewing these data years after the project was active, it occurred to me that the problem of lowering the cure temperature could also have been approached by using only PACM as a curative, and making the necessary adjustments for exotherm and viscosity by finding the right blend of diepoxide diluent with monoepoxide diluent. Apparently, the presence of monoepoxide has a huge effect on reactivity. The monoepoxide butyl glycidyl ether (BGE) is the one used to reduce the viscosity of Epon 828. It is 11% of Epon 815 resin. BGE and most other monoepoxides are more mobile than their diepoxide cousins and get to react with the most reactive amine groups first. Heat is liberated from their reaction, which in turn accelerates the diepoxide reactions. Small diepoxide molecules, like Heloxy 67, are more mobile than the DGEBA molecules in Epon 828, but once they react, the second epoxide on the Heloxy 67 becomes fairly immobile. At any rate, the test data indicate that the differences in peak exotherm between using monoepoxides and small diepoxides are sizeable. It could be possible to blend them to obtain the desired compromise.

KEYWORDS

- encapsulation
- epoxy
- steric hindrance
- exotherm test
- cyanate ester

CHAPTER 9

PLATABLE ADHESIVES FOR CYANATE ESTER COMPOSITES

CONTENTS

9.1   BACKGROUND

9.1.1   A FIBER-PLASTIC COMPOSITE, MICROWAVE CHANNELING DEVICE

One of the most important goals in designing spacecraft is minimizing weight. One way to make big weight savings is to replace metal parts with lighter plastics or composite parts. For microwave channeling devices, the wave guide needs to be metal. However, the bulk of the device could be a reinforced plastic as long as there is metal on the surface of the part. In principle, one could replace an all-metal part with a composite part that has metallic plating on its surfaces.

Brian Punsly, of Hughes Space and Communications Group, had the task of making that concept a reality. We shall be discussing work that led to a patent for a platable adhesive in this chapter. However, this patent is only one of the family of patents granted to him. One of other patents was entitled, "Preparation of Cyanate Ester Polymers and Composites for Metal Plating". This new process is highly important to the development of a platable adhesive.

Cyanate ester polymers and graphite fibers comprise a composite material with excellent dimensional stability over a range of temperatures and humidities. These properties are essential in microwave channeling devices.

Cyanate ester polymers are high temperature, thermoset polymers having distinct advantages for electronic applications. Principally, they are used as the matrix materials for printed wiring board laminates and structural composites. They have low dielectric loss characteristics and dimensional stability at elevated temperatures. They are tough and resistant to micro-cracking caused by thermal cycling.

9.1.2   PLATING ON PLASTICS TECHNOLOGY

Don Helber was TSD's plating expert. He had many years of experience and had once run his own plating company. Don, with his plating experience, and I, with my plastics background, teamed up to offer our services to Brian Punsly, who had the assignment to find a method to silver plate on graphite fiber-reinforced cyanate ester composites. Our Malibu Research Laboratories were also vying for this development work.

Don Helber put his heart and soul into this project and ended up in a major development, which resulted in a successful process to prepare the composite material to be plated. I picked up assignments to facilitate this effort too. I obtained different etchants to treat the composite surfaces in order to get good adhesion of the plating to the composite. Although this effort would make an interesting story, it is only peripheral to the platable adhesive development story, which led to a patent. So, let us move on to the platable adhesive development problem.

9.2   THE PROBLEM

The assembly consisted of smaller cyanate ester composite parts bonded together using an epoxy adhesive. The assembled structure was then etched to enhance plating adhesion and silver plated. The problem was that the plating did not adhere to the adhesive bond line. A platable adhesive was needed to solve this problem. The desired adhesive needs to cure at room temperature or slightly above. It has been determined that higher temperatures would warp the structure. It will provide a strong bond between cyanate ester composite parts. The cured adhesive will also be rendered platable by the same process used on the cyanate ester composite parts.

9.3   APPROACH

We had a large investment in learning how to prepare cyanate ester composites for plating. I reasoned that if we could formulate an adhesive filled with cyanate ester powder, the same etching process would work on the adhesive, which worked so well on cyanate ester composites. Moreover, the etched adhesive would provide anchoring sites to more strongly hold the plating. The etchant removes the cyanate ester at a faster rate than it removes the epoxy matrix of the adhesive. These crevices provide a mechanical lock, on which the plating can adhere.

9.4   EXPERIMENTATION

Wai Chen Seetoo had joined our staff around this time. She did the majority of the laboratory work and testing. She focused mainly on finding an adhesive which could pass the 'Plating Adhesion Test.' Of most concern to us is that the etched adhesive can be plated with our plating process. Test adhesive specimens were etched using the same process that we had optimized for the cyanate ester composites. A layer of copper was deposited followed by a layer of silver. The adhesion of the plating was evaluated using the following test method:

9.4.1   ADHESION TESTING OF PLATED SPECIMENS

Plated specimens were tested for how well the plating adhered to the adhesive candidates by a modified version of ASTM D3359, "Test Method B - Cross-Cut Tape Test." An area 1 inch by 1 inch is scribed by cutting through the plating with a razor-sharp blade to produce a grid pattern. Calibrated tape is applied to the test area and peeled off. A numerical rating is assigned to the specimen depending on how many squares of plating remain. We scored the cross-cut test as follows: if the tape was unable to remove any plating, we scored it a 5 points. If it removed all the plating, we scored it zero points. The intermediate scores were 1, 2, 3, or 4 points as per the criteria of Table 9.1. Our modification of ASTM D3359 entailed using more aggressive tapes than the standard required. Instead of limiting our testing to 45 oz/per inch width tape, we used up to 150 oz per inch width tape.

TABLE 9.1   Tape Adhesion Test Numerical Rating System

Rating	Criteria
5	Perfect adhesion
4	Slight delamination alone edges of the cut
3	Surface area equal to one full square removed
2	Two squares of plating removed
1	Three squares of plating removed
0	More than half of cross-hatch region removed

9.4.2   TEST RESULTS FOR UNFILLED EPOXY ADHESIVES

The previous effort on the platable adhesive had focused on evaluating epoxy adhesives currently used at Hughes Aircraft Company. In all, 79 different epoxy adhesives had been tested. Epiphen 825A, B.F. Goodrich

1177, B.F. Goodrich 1273, Hysol 9395, and Hysol 9394 are a few of them. This Edisonian approach to the problem failed. None of them were platable.

9.4.3 TEST RESULTS FOR CYANATE ESTER ADHESIVES

Cyanate esters had also been evaluated as adhesives for the composite parts. The following table is taken from U.S. Patent 5,840,829. I am showing it to illustrate that finding an acceptable adhesive for this application is not apparent to someone skilled in the art.

TABLE 9.2 Results for Cyanate Ester Adhesives

Trial #	Adhesive designation	Total Cure	Tackiness	Plating texture	Tape test 135 oz	Tape test 150 oz
1	Bryte 083194-1M	200°F for 1.5 h	Tacky	4	3+	0
2	083194-1M + 10% catalyst 119	200°F for 1.5 h	Slightly tacky	4	5	0
3	083194-1M + 20% catalyst 119	200°F for 1.5 h	Slightly tacky	3	3	0
4	Bryte 083194-1M	175°F for 3.5 h	Tacky	4	4+	0
5	Bryte 083194-1M	130°F for 15 h	Tacky	4	0	0
6	EX-1502	250°F for 1.5 h	Tacky	4	0	0
7	083194-1M + 20% catalyst 119	200°F for 3.5 h	Slightly tacky	4	4	0
8	EX-1502	250°F for 3.5 h	Tacky	4	-	4
9	Bryte 083194-1M	200°F for 3.5 h	Tacky	3-	-	1

The plating adhesion to the test adhesives was inferior to what is attainable to the cyanate ester composites. The problem is that the cyanate esters in the adhesives require temperatures much higher than the 120°C

(248°F) at which the composite parts warp. Cyanate esters are normally cured at 250–350°C (482–662°F) for several hours. Therefore, the cyanate ester adhesives in the table above were too under-cured to provide good plating adhesion.

9.4.4 EPOXY ADHESIVES CONTAINING CYANATE ESTER POWDER

Now, we discuss an approach that worked extremely well in contrast to the two previous approaches. I needed the low temperature cure characteristics of an epoxy resin/aliphatic polyamine formulation, but I wanted the etchant to act on the cured adhesive as though it was a cured cyanate ester. The key to accomplishing this was to grind a cured cyanate ester to a fine powder and stir it into the liquid epoxy adhesive. We wanted to maximize the amount of cyanate ester powder in the adhesive, but not to the point where adhesive strength suffered.

9.4.4.1 TEPA VERSUS TETA

The TEPA curative seemed superior to TETA curative in obtaining a tack-free surface. The structures are shown below:

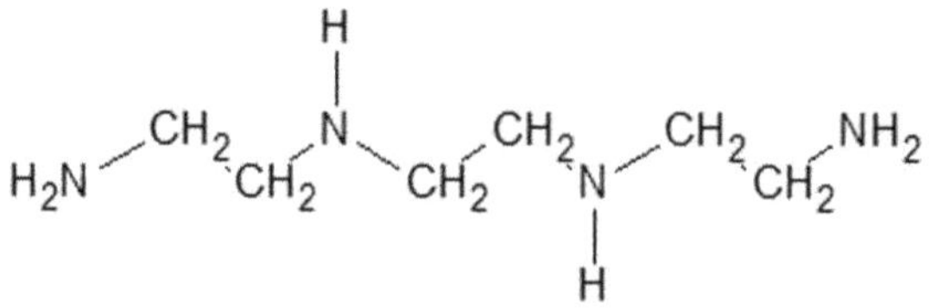

FIGURE 9.1 Structure of TETA curative.

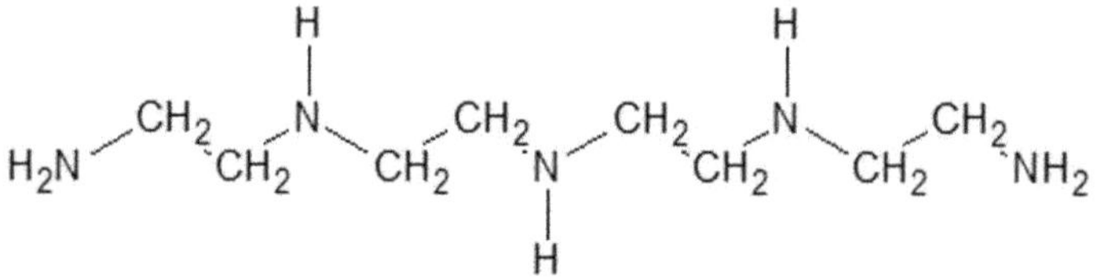

FIGURE 9.2 Structure of TEPA curative.

9.4.4.2 TEST RESULTS

The following table shows the epoxy adhesive formulations which were filled with different cyanate ester powders and cured under different conditions. The texture and tape test values per ASTM D3359 for the different trials are also listed. These adhesives performed so well for providing good plating adhesion that only the most aggressive tape (150 oz.) was used during our testing.

TABLE 9.3 Results on Cyanate Ester Powder-Filled Epoxy Adhesive Formulations

Trial #	Formulation, (epoxy resin, curative, cyanate ester powder)	Powder size, microns	Cure condition	Tackiness	Texture	Tape test, 150 oz
1	Epon 815 - 150 pbw TEPA - 18 pbw BtCy-1 - 50 pbw	25	20 days, against EF-179 coated steel	Non-tacky laminated	5	5-
2	Epon 815 - 150 pbw TEPA - 18 pbw CE 083194-1M - 50 pbw	25	20 days, against EF-179 coated steel	Non-tacky laminated	4	5
3	Epon 815 - 150 pbw TEPA - 18 pbw EX-1515 - 50 pbw	25	20 days, against EF-179 coated steel	Non-tacky laminated	4	4+
4	Epon 828 - 150 pbw TEPA - 18 pbw BtCy-1 - 50 pbw	25	20 days, against EF-179 coated steel	Non-tacky laminated	3	5
5	Epon 828 - 150 pbw TEPA - 18 pbw BtCy-1 - 50 pbw	25	21 days, exposed in air	Slightly tacky	4	5
6	Epon 828 - 150 pbw TEPA - 18 pbw CE 083194-1M 50 pbw	25	21 days, exposed in air	Tacky	3	5
7	Epon 828 - 150 pbw TEPA - 18 pbw EX-1515 - 50 pbw	25	21 days, exposed in air	Very tacky	4	5

TABLE 9.3 *(Continued)*

Trial #	Formulation, (epoxy resin, curative, cyanate ester powder)	Powder size, microns	Cure condition	Tackiness	Texture	Tape test, 150 oz
8	Epon 815- 150 pbw TEPA - 18 pbw BtCy-1 - 50 pbw	 25	6 days, exposed in air	Tacky	4	0
9	Epon 815- 150 pbw TEPA - 18 pbw BtCy-1 - 50 pbw	 25	6 days, against Teflon	Non-tacky laminated	4	0
10	Epon 815- 150 pbw TEPA - 18 pbw EX-1515 - 50 pbw	 25	7 days at RT, 150°F for 2 h	Slightly tacky	4	0
11	Epon 815- 150 pbw TETA - 18 pbw CE 083194-1M 50 pbw	 25	8 days, exposed in air	tacky	4	5
12	Epon 815- 150 pbw TETA - 18 pbw CE 083194-1M 50 pbw	 25	8 days, against Teflon	Non-tacky laminated	4	0
13	Epon 815- 150 pbw TETA - 18 pbw CE 083194-1M 50 pbw	 25	7 days at RT, 150°F for 2 h	Slightly tacky	3	2
14	Epon 815- 150 pbw TETA - 18 pbw BtCy-1 - 50 pbw	 25	7 days at RT, 150°F for 2 h	Slightly tacky	3	0
15	Epon 815- 150 pbw TETA - 18 pbw EX-1515 - 50 pbw	 25	5 days, exposed in air	Tacky	4	5
16	Epon 815- 150 pbw TETA - 18 pbw EX-1515 - 50 pbw	 25	8 days, against Teflon	Non-tacky laminated	4	3
17	Epon 815- 200 pbw TETA - 24 pbw CE 083194-1M 100 pbw	 45	4 days in air,	Tacky	3	4

TABLE 9.3 *(Continued)*

Trial #	Formulation, (epoxy resin, curative, cyanate ester powder)	Powder size, microns	Cure condition	Tackiness	Tex-ture	Tape test, 150 oz
18	Epon 815- 50 pbw Epon 828 - 50 pbw TETA - 12 pbw EX-1515 - 50 pbw	45	5 days, exposed in air	Tacky	3-	2
19	Epon 815- 50 pbw Epon 828 - 50 pbw TETA - 12 pbw EX-1515 - 50 pbw	45	4 days, against Teflon	Non-tacky laminated	2	3-
20	Epon 815- 50 pbw Epon 828 - 50 pbw TETA - 12 pbw BtCy-1 - 50 pbw	45	4 days, exposed in air	Tacky	4	4+
21	Epon 815- 50 pbw Epon 828 - 50 pbw TETA - 12 pbw BtCy-1 - 50 pbw	45	4 days, against Teflon	Non-tacky laminated	3	1
22	B.F. Goodrich 1177 A & B - 100 pbw EX-1502 - 50 pbw	45	7 days, exposed in air	Slightly tacky	5	5
23	B.F. Goodrich 1177 A & B - 100 pbw EX-1502 - 50 pbw	45	4 days, against Teflon	Non-tacky laminated	5	0
24	B.F. Goodrich 1177 A & B - 100 pbw BtCy-1 - 50 pbw	45	7 days, exposed in air	Slightly tacky	5	5

Comments:

1. In general, the adhesives filled with cyanate ester powder have ad-hesion of rating level 5 when tested with the most aggressive tape

(i.e., 150 oz/inch width). One exception is TETA or TEPA curative-containing formulation, which were cured against Teflon.

2. Twenty-five (25) micron particle size powder yielded better plating adhesion than 45 micron particle size. Room temperature cures yielded better plating adhesion than oven cures.

3. The type of cyanate ester powder made a difference too. From best to worst, the order is BTCy-1, E083194-1M, EX-1502, and EX-1515.

9.5   THE SOLUTION

The successful adhesive formulation consists of a liquid epoxy resin, stoichiometric amounts of an aliphatic polyamine curing agents, and 15–50% by volume of powdered cyanate ester plastic.

9.5.1   THE ADHESIVE INGREDIENTS

Table 9.4 below summarizes the formulation information for a successful platable adhesive.

TABLE 9.4   Successful Platable Adhesive Formulation

Function	Favored Ingredient	Comment
Epoxy resin	Epon 815 and Epon 828	
Curative	TEPA, tetra-ethylene pentamine	Less tacky surface
Filler	Cured cyanate ester, ground to a powder, BTCy-1 preferred	25 microns max., 15–50 wt.% of formula

9.5.2   KEY TO THE ADHESIVE PREPARATION PROCESS

The following lessons were learned:

1. The cyanate ester filler is best incorporated into the epoxy resin/curative mixture using a three-roll mill. The resulting adhesive has the consistency of a flowable paste.

2. Stirred-in air can be eliminated by mixing the ingredients under vacuum.
3. The adhesive is packaged in Semco cartridges and frozen as a frozen premix adhesive.

9.5.3 APPLICATION OF THE ADHESIVE

The platable adhesive is processed as follows:

1. Thaw the frozen premix adhesive.
2. Apply adhesive to cyanate ester composite part surface to form a bondline 0.002–0.010 inch thick.
3. Use fixtures to hold aligned parts in position during the cure cycle. Apply modest pressure to the bondline during cure schedule.
4. Cure at room temperature for best plating adhesion results.

9.6 PATENTS

U.S. Patent 5,780,581 titled, "Platable Structural Adhesives for Cyanate Ester Composites" was awarded to Ralph D. Hermansen, Brian M. Punsley, and Wai Cheng Seetoo on October 10, 1998. The application was filed October 27, 1995.

The above patent is superseded by U.S. Patent U.S. 5,840,829 having the same title and same assignees as U.S. Patent 5,780,581. The latter patent was awarded on November 24, 1998 and had been filed for on November 25, 1997. There is very little difference between the two patents.

KEYWORDS

- cyanate ester polymers
- platable adhesive
- plating
- epoxy adhesives

CHAPTER 10

REWORKABLE, THERMALLY CONDUCTIVE, ADHESIVES FOR ELECTRONIC ASSEMBLIES

CONTENTS

10.1   THE ENGINEERING PROBLEM

This project had to do with a bid on a Naval weapons program, so I am not going to elaborate on what the application was, but limit my discussion to the technical problem of finding a material and process solution to a set of engineering objectives. Our assignment was to investigate the feasibility of solving a heat conduction problem as part of a bid to the navy for a new weapon system.

Richard F. Davis needed a material and process to bond heat sinks to a printed wiring assembly. The material would as a thermally conductive path for electronic components to transfer heat to a metallic heat sink. Rick needed the best thermal conduction possible while meeting the other requirements. Values of 2–3 BTU/h-ft-°F were sought. The technical problem is to bond a heat sink to a printed wiring assembly (PWA) using an adhesive which is a dielectric as well as a thermally conductive material. The process must not cause dislocation of electronic components nor breakage of any circuitry. The resultant bond should be strong enough to hold the heat sink in place, even during mechanical shock waves, but weak enough to allow easy removal of the heat sink without damage to components, if removal is deemed necessary.

Moreover, the adhesive must be flexible enough to accommodate thermal stresses from joined materials having different coefficients of thermal expansion. Removal of the bonded heat sink without damage to components is a difficult requirement because the greater the surface area of a heat sink, the more difficult it is to remove. Therefore, the adhesive needed to be very soft and low in modulus. A Shore A durometer under 35 was deemed necessary. Temperatures as low as 5°C will be experienced in the application environment.

10.2   TRANSLATING APPLICATION OBJECTIVES INTO MATERIAL AND PROCESS CRITERIA

Table 10.1 lists the objectives for the application and translates those objectives into a description of a complying material and what properties that material must have.

TABLE 10.1   Translation of Application Objectives into M & P Criteria

Objective	Material description	Properties
Maximized thermal transfer, 2–3 BTU/h-ft-°F	Highly filled elastomer	Pre-filled compound ruled out. That leaves filler first and impregnate filler with polymer
Good dielectric properties	Dielectric filler in good dielectric polymer	Hydrophobic polymer best
Controlled strength a) Shock resistant and durable b) Reworkable	The cured polymer has to be very soft to be both reworkable and absorbing energy in thin cross-sections. It must be tough enough to survive application needs	Shore A durometer of about 30, Tg < 5°C Flexible epoxy for good adhesion to metal, etc.
Amenable for introducing thermal transfer adhesive on hardware	Impregnating small filler particles means very low viscosity mixed polymer having long work-life	Mixed A & B < 1000 cps Time to 2× viscosity > 60 min But low temperature cure for electronics

10.3   OUR FAVORED APPROACH—IMPREGNATION OF A CONDUCTIVE POWDER

The thermally conductive adhesives, which we encountered in Part 2 of the book, were all particulate-filled compounds. These paste adhesive compounds were limited in thermal conductivity (0.2–0.5 BTU/h-ft-°F) because they had to be flowable liquids in order to be applied. Now, we are attempting to maximize thermal conductivity (2–3 BTU/h-ft-°F was sought). We can approach that goal by:

(1) Sealing the periphery of the gap between circuitry and heat sink
(2) Introducing a conductive powder filler into the cavity between circuitry and heat sink
(3) Vibrating to maximize packing of filler
(4) Using a flexible epoxy impregnant to fill the voids between the particles
(5) Curing the impregnant

10.4   PRELIMINARY INVESTIGATION

10.4.1   FILLER SELECTION

Part of the task is to select the best filler for the application. Alumina and other conductive fillers are candidates. The particle size of the filler is limited at the high end by the bondline thickness. Ten mil bondlines were a possibility. Thermal conductivity data from previous testing were compiled in Table II-8-2 to demonstrate that the goal is attainable.

TABLE 10.2   Thermal Conductivity Results for Different Filler Mixtures

Alumina	Aluminum nitride	Boron nitride	% Filler volume in adhesive	Thermal Conductivity, BTU/h-ft-°F
80%	None	20 %	62.7	2.3
69%	31 %	None	68	3.5
100%*	None	None	73	2.5
69%	31 %	None	62	2.1

*A mixture of different sized alumina particles to maximize packing.

10.4.2   PROPERTY CRITERIA FOR THE IMPREGNANT

The major part of the task is to define and formulate a suitable impregnant. A tentative list of property requirements was compiled for the new impregnant based on the objectives for the application. The following table shows them:

TABLE 10.3   Tentative Requirements for the Flexible Epoxy Impregnant

Property	Test method	Requirement
Viscosity (penetration ability)	Brookfield viscosity	Less than 1000 cps
Work-life (penetration time)	Brookfield viscosity	Time to double viscosity > 1 h
Durometer, Shore A (shock resistant, reworkable)	ASTM D 2240	35 or lower
Volume resistivity	ASTM D 257	Above x 10^9 Ohm-cm
Dielectric constant	ASTM D150	Below 6.0
Glass transition	Various	under 5°C

10.5   FORMULATION DEVELOPMENT AND TESTING OF THE IMPREGNANT

10.5.1   FLEXIBLE EPOXY FORMULATIONS FOR THE IMPREGNANT

E. Dean Johnston did the laboratory work on this project and was a co-patentee. Steve Lau advised on formulating and impregnation. Table 10.4 shows the formulation, Durometer, viscosity, and change in viscosity with time for three successful candidates. A fourth candidate, Heloxy 84 and Dytek A, did not properly cure and is omitted here.

TABLE 10.4   Formulation and Properties of Flexible Impregnant Trials

Trial number	Formulation	Shore A, initial	Shore A, heat aged	Viscosity, initial	Viscosity, 1 h later	Viscosity, 2 h later
1	Epon 871 100 pbw Dytek A 6,8 pbw	35	69	400	470	500
2	Heloxy 505 100 pbw DP-3680 28 pbw	20	33	600	900	1550
3	Heloxy 505 100 pbw Dytek A 4.8 pbw	18	32	550	800	1150

Chemical structures for Heloxy 505 and Dytek A are shown below:

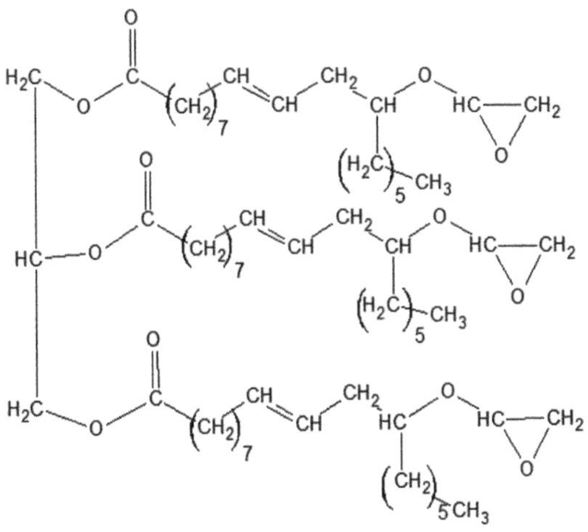

FIGURE 10.1   Structure for Heloxy 505 epoxy resin.

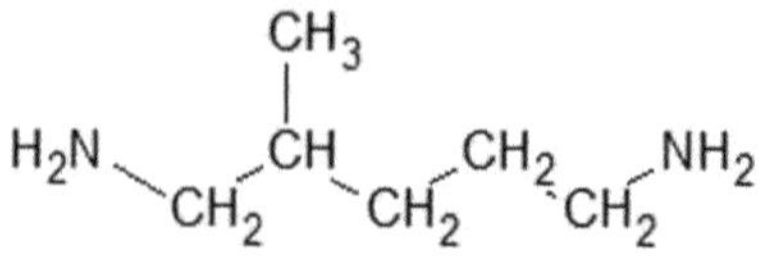

FIGURE 10.2   Structure for Dytek A curative.

10.5.2 MOCK UP IMPREGNATION

A mock up simulation of the impregnation was conducted by using a 4 inch × 4 inch square aluminum plate for the heat sink and a similar-sized glass plate for the PWB. The two items were separated by 90 mils. The cavity was filled with 30 mil glass microspheres. The assembly was impregnated using Trial # 2 impregnant. The filler was impregnated in only a few minutes.

10.5.3 IMPREGNATION OF DIFFERENT GAP THICKNESSES

In order to simulate a populated circuit board containing flatpacks, a metal plate was machined having twenty-five (25) 1 inch × 1 inch squares. We also used the opportunity to study different bondlines or gaps for the transfer adhesive to fill. Consequently, the squares were machined to provide gaps of 5, 10, 20, 30, and 40 mils. A glass plate simulated the PWB and provided a means of visually inspecting the addition of fillers and quality of impregnation. Three different sizes of spherical fillers were evaluated for filling the gaps. The fixture was vibrated to attain maximum filler packing. The assembly was impregnated using DER 732 epoxy resin without curative. We wished to reuse the fixture as often as needed so we did not want to cure the impregnant. Ways of improving the impregnation

TABLE 10.5   Results of Filling and Impregnating Different Size Gaps

Filler Name	Particle size	Gaps not filled with particles	Impregnation time
Glass spheres	8–10 mil	5 and 10 mil	No test
Potter's 2227 CPO3	115 microns	5 mil	2.5 h with vacuum*
Potter's 2530 CPO3	65 microns	none	3.0 h with vacuum

*Impregnation took 5 h without vacuum assist.

were learned with each progressive round of testing. The following table summarizes our findings:

10.6 CONCLUSION

This project was halted suddenly and the funding was cut off. This is a common experience when a bid is being prepared, because there are always limited budgets for only tentative work. Fortunately, we had proven the feasibility of the concept before funding ended. I wrote a progress report to show that we had a solution to the problem and to document the details, in case it got started up again. Rich Davis was interested in applying for a patent on our invention so that part of the effort went forward.

10.7 PATENTS

U.S. Patent 6,132,850, titled "Reworkable, Thermally-Conductive Adhesives" was awarded to Ralph D. Hermansen, R.F. Davis, and E.D. Johnston on October 17, 2000. The assignee was now Raytheon Company, who had purchased the military part of Hughes Electronics Corporation. I had retired already when this patent was granted. The patent application was filed on November 25, 1996, when I still worked for Hughes.

The patent title is a little misleading, because the patent is really about an impregnant and that word probably should have been in the title.

KEYWORDS

- electronic components
- adhesives
- thermal conductivity

CHAPTER 11

ROOM TEMPERATURE-STABLE, ONE-COMPONENT, FLEXIBLE EPOXY ADHESIVES

CONTENTS

11.1   THE OPPORTUNITY

During the process of doing formulation development work for Delco Electronics Corporation, Steve Lau and I discovered that our Flexipoxy formulations could be made into room temperature-stable compounds, which would adequately cure at low enough temperatures for electronic assembly purposes. If we could apply these lessons learned to Hughes applications, we could offer compounds, which no longer need to be stored in frozen form, but could be stored at room temperature. Moreover, it would be an advantage to manufacturing to avoid the thawing of a frozen premix to room temperature, but be able to use the product immediately.

Currently, the Flexipoxy compounds are available as two-component kits or one-component frozen premix cartridges. It is preferable not to have to weigh and mix ingredients due to possible weighing errors or under-mixing problems. The two components can be precisely weighed and properly mixed by specialists, then quick frozen as single-component systems. The next improvement would be to remove the necessity for freezer storage entirely.

11.2   THE PROJECT

In 1992, Steve Lau and I were granted modest funding from our division to research and develop one-component, room temperature-stable, flexible epoxy (1C-RTS-FE abbreviated) compounds, and seek to patent them. Part of the early effort involved literature searches and vendor searches for applicable latent curatives.

The focus of the formulation effort was to develop three different 1C-RTS-FE adhesives, namely:

(1) A 1C-RTS-FE general-purpose adhesive,
(2) A 1C-RTS-FE thermal transfer adhesive,
(3) A 1C-RTS-FE electrically conductive adhesive.

11.3   FORMULATION GOALS FOR THE THREE ADHESIVES

Based on our previous experience of developing adhesives with these purposes, we compiled a list of desired properties for each of the three 1C-RTS-FE adhesives. The following three tables show those objectives:

TABLE 11.1   Property **Goals for a 1C-RTS-FE General-Purpose Adhesive**

Property	Test method	Requirement
Storage life at room temperature	ASTM D2240	3 months minimum
Adequate cure at 100–125°C	ASTM D2240	2 h or less
Shore A durometer	ASTM D2240	95 A maximum
Ultimate elongation in tension, %	ASTM D412	80% minimum
Lap shear strength (alum–alum)	ASTM D1002	250 psi minimum

TABLE 11.2   Property Goals for a 1C-RTS-FE Thermally Conductive Adhesive

Property	Test method	Requirement
Storage life at room temperature	ASTM D2240	3 months minimum
Adequate cure at 100–140°C	ASTM D2240	1 h or less
Shore A durometer	ASTM D2240	90 A maximum
Thermal conductivity	Colora or equivalent	4.0 BTU/h-ft-°F minimum
Volume resistivity,	ASTM D257	
(1) At room temperature		1.0×10^{12} ohm-cm minimum
(2) At 93°C		1.0×10^{10} ohm-cm minimum
Lap shear strength (alum–alum)	ASTM D1002	250 psi minimum

TABLE 11.3   Property Goals for a 1C-RTS-FE Electrically Conductive Adhesive

Property	Test method	Requirement
Storage life at room temperature	ASTM D2240	3 months minimum
Adequate cure at 100–140°C	ASTM D2240	2 h or less
Shore A durometer	ASTM D2240	95 A maximum
Glass transition temperature	ASTM E831	Lower than 10°C
Volume resistivity	ASTM D257	1.0×10^{-2} ohm-cm maximum
Lap shear strength (alum–alum)	ASTM D1002	
(1) 1 h cure at 120°C		100 psi minimum
(2) 2 h cure at 120°C		150 psi minimum

11.4   GENERAL APPROACH FOR 1C-RTS-FE ADHESIVES

One-component epoxy compounds, stable at room temperature, already existing for hard, rigid epoxies. Their formulators simply substituted a latent curative for an active curative in the formulation. What was un-known is whether latent curatives would also work on flexible epoxy res-ins. We needed to determine if such compounds would be stable at room temperature for months and whether the cured compound would be as flexible as required.

Note: Many latent curatives are solids at room temperature. If they are ground to a fine powder and stirred into a liquid epoxy resin, negligible reaction takes place until that mixture is raised to the melting point of the curative.

Our approach was to combine different flexible epoxy resins with la-tent curatives and determine if they are compatible, cure within an hour at temperatures under 140°C, and are flexible once cured. Flexibility is determined by measuring the Shore A durometer and tensile ultimate elon-gation. We determined whether or not the trial compound was a suitable adhesive by preparing and testing lap shear specimens. These specimens were aluminum bonded to aluminum.

11.4.1 THEORETICAL CONSIDERATIONS

In the case of liquid epoxy resins and liquid curatives, it is understood that polymerization is a chemical reaction and that a very intimate mixing of the two reactants is desirable for a complete reaction. However, in the case of a solid curative in powder form stirred into a liquid epoxy resin, we do not have the same intimate mixing of reactants. Remember that there is an Avogadro's Number of molecules (6.02×10^{23}) in the gram molecular weight of any molecule, so there may be trillions of curative molecules in each tiny solid curative particle.

The important point is that the finer the particle size of the solid cura-tive, the better the intimacy of mixing of co-reactants because there are fewer curative moles per particle. Since the volume of a particle is a cu-bic function of the lineal size, there is an eightfold decrease in curative molecules when we halve the particle size. Moreover, halving the particle size increases the total surface area of the curative particles fourfold, thus providing more immediate contact area for reaction.

Due to reaction on the particle's surface, a micro-thin coating of react-ed epoxy eventually surrounds each curative particle. This coating bursts when the particle melts and expands. The hot liquid curative molecules mix with the epoxy resin molecules due to thermal effects. The higher the selected cure temperature, the better the dispersion of curative molecules into the surrounding epoxy molecules.

Adhesives, containing mineral or metal particles, are especially interesting because the solid curative particles compete for space with the filler particles. Ideally, the curative particles could fit into the voids between filler particles if the relative sizes allow it. At the cure temperature, the situation changes as the curative liquefies. Finally, after polymerization, the filled adhesive is a two-phase system of cured polymer and filler.

11.4.2 APPROACH FOR DEVELOPING THE 1C-RTS-FE GENERAL-PURPOSE ADHESIVE

Our approach was to combine different flexible epoxy resins with latent curatives and determine if they are compatible, cure within an hour at temperature under 140°C, and are flexible once cured. Flexibility was determined by measuring the Shore A durometer and tensile ultimate elongation. Whether the compound is a suitable adhesive was determined by bonding aluminum to aluminum; lap shear specimens determine the lap shear strength of the cured specimens.

11.4.3 APPROACH FOR DEVELOPING THE 1C-RTS-FE THERMAL TRANSFER ADHESIVE

Alumina powder has been our traditional filler of choice to increase the thermal conductivity of an adhesive, while also attaining good dielectric properties. We had to consider that both the latent curative and the alumina powder are solids and thus compete for the available physical space. It is best if the alumina particles touch each other in order to maximize thermal conductivity. The only available space for the curative is in the interstices of the alumina particles. A latent curative must have sufficiently fine particle size and be effective at low enough percentage of the flexible epoxy resins.

The latent curative will liquefy at the cure temperature leaving the alumina as the only solid during that stage in the process.

11.4.4   APPROACH FOR DEVELOPING THE 1C-RTS-FE ELECTRICALLY CONDUCTIVE ADHESIVE

Electrically conductive adhesives can be formulated by adding a metal powder to a liquid mixture of epoxy resin plus curative. When the volume fraction of metal reaches a point where the metal particles touch each other, the adhesive becomes electrically conductive. Metals of choice are silver, gold, platinum, and other non-corroding metals. From a cost standpoint silver is the most practical to use.

The biggest difficulty is that the latent curative and the silver flake are both solids and thus compete for the available physical space. However, it is vital that the silver filler particles touch in order that electrical conductivity is attained. A latent curative must have sufficiently fine particle size and be a minimal percentage of the formulation so that the electrically conductive filler can dominate. The metallic filler must occupy about 40% of the volume of the adhesive to be electrically conductive. We need to find epoxy resin/latent curative combinations with maximum liquid percentage.

11.5   FORMULATION DEVELOPMENT OF 1C-RTS-FE GENERAL-PURPOSE ADHESIVES

11.5.1   LATENT CURATIVES

Additional latent curatives were included in the formulation development of the 1C-RTS-FE thermal transfer adhesives. Table 11.4 shows their trade names and properties.

TABLE 11.4   Description of Latent Curatives

Latent curative	Chemical description	Parts/100 parts resin	Particle size	Melting Point
Ajicure AH-122	Aliphatic dihydrazide	58		120°C
Ajicure PN-23	An amine adduct	20	10–12 microns	100°C
Ajicure AH-127	4-Isopropyl-2,5-dioxoimidazolidine-1,3-di (propionohydrazide)	25		115°C
Ancamine 2014FG	modified polyamine	28	Under 6 microns	98–106°C

Comments:

1. The Ajicure products are made by Ajinomoto, and Ancamine is made by Air Products Company. The following figure shows the structure for a hydrazide:

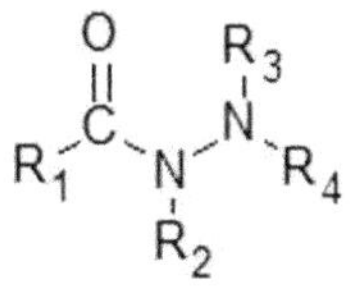

FIGURE 11.1   A hydrazide.

2. Ajicure AH-127 adds flexibility to the epoxy product.

11.5.2   SCREENING OF LATENT CURATIVES

Table 11.5 summarizes the screening of three latent curatives using four different epoxy resins.

TABLE 11.5   Performance of Different Epoxy Resins with Different Latent Curatives

Epoxy resin	Cure Temperature	Ancamine 2014-FG	Ajicure PN-23	Ajicure AH-122
Epon 828	80°C	Rigid in 1 h	No test	No test
Epon 828	100°C	Rigid in 30 min	Rigid in 30 min	Rigid in 1 h
Epon 828	120°C	No test	Rigid in 15 min	Rigid in 30 min
Cardolite NC-514	100°C	Did not cure	Rigid in 1 h	Flexible in 1 h
Cardolite NC-514	120°C	Rigid in 2+ h	Rigid in 30 min	Flexible in 0.5 h
Epon 871	100°C	Did not cure	Incompatible	Flexible in 3+ h
Epon 871	120°C	Flexible in 2+ h	Incompatible	Flexible in 2 h
Cardolite NC-547	100°C	Did not cure	Incompatible	Flexible in 2 h
Cardolite NC-547	120°C	Flexible in 2 h	Incompatible	Flexible in 1 h

Comments:

1. Epon 828 was known to produce rigid epoxy plastics with these latent curatives.

2. Cardolite NC-514 produced rigid plastics with 2014-FG and with PN-23, but produced a flexible material with AH-122.
3. Epon 871 produced a flexible material with 2014-FG and with AH-122, but was incompatible with PN-23.
4. Cardolite NC-547 produced a flexible material with 2014-FG and with AH-122, but was incompatible with PN-23.
5. Ajicure PN-23 was ruled out as a curative for making a flexible epoxy.
6. Ancamine 2014-FG and Ajicure AH-122 both produced flexible epoxies, but AH-122 worked at lower cure temperatures. Moreover, AH-122 produced a flexible epoxy with Cardolite NC-514 whereas it was rigid with 2014-FG.
7. Ajicure AH-122 is our first choice for formulating flexible epoxies.

11.5.3 FORMULATING 1C-RTS-FE GENERAL-PURPOSE ADHESIVES

By utilizing Ajicure AH-122, we did formulate five successful candidates. The formulations and properties are shown in the table below:

TABLE 11.6 Formulations and Properties of 1C-RTS-FE General-Purpose Adhesives

Flexipoxy designation	Formulation	Durometer, Shore A	Ultimate elongation, %	Lap shear strength, psi
RTS-1000	100 pbw Heloxy 67	88	110	870
	100 pbw Ajicure AH-122			
RTS-1100	50 pbw Heloxy 67	45	90	120
	50 pbw Heloxy 84			
	60 pbw Ajicure AH-122			
RTS-1200	50 pbw Heloxy 67	42	105	385
	50 pbw DER 736			
	85 pbw Ajicure AH-122			
RTS-1300	100 pbw DER 736	36	110	190
	71 pbw Ajicure AH-122			
RTS-1400	100 pbw DER 732	15	250	40
	42 pbw Ajicure AH-122			

Comments:

1. Heloxy 84 epoxy resin has the following structure:

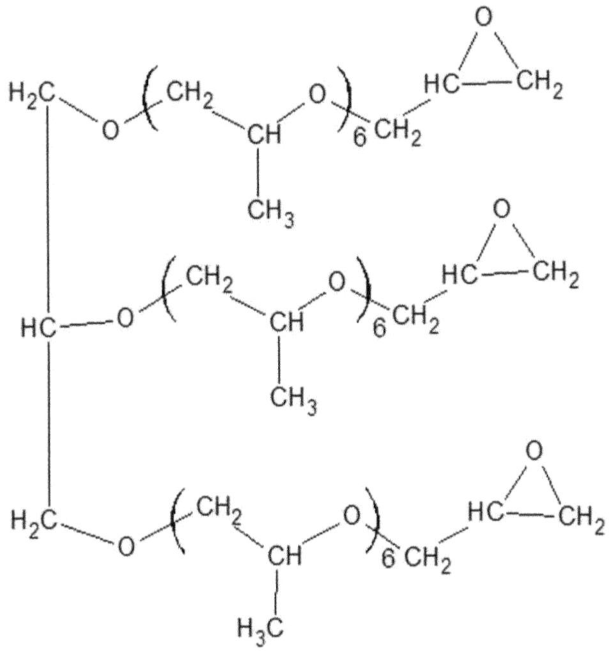

FIGURE 11.2   Heloxy 84 epoxy resin.

2. One particular latent curative, Ajicure AH-122, was useful with different flexible epoxy resins to produce a wide spectrum of flexibilities. Ajicure AH-122 is a dihydrazide having an active hydrogen equivalent weight of 134.

3. Durometers from Shore A 15 through to Shore A 88 were obtained. Moreover, a 1C-RTS-FE adhesive of any intermediate Shore A durometer could be obtained by blending the above formulations. For example, a Shore A Durometer of 60 should be easily obtainable by blending RTS-200 and RTS-201 together.

4. The ultimate tensile elongations were all very close to 100% except for RTS-40, which was a very soft elastomer.

5. On the other hand, the adhesion of the aluminum-to-aluminum lap shear specimens varied widely from formulation to formulation. The weaker adhesives might be best for applications where

reworkability is important. The stronger adhesives might be best where reliability of the bond is most important.

11.6   FORMULATION DEVELOPMENT OF 1C-RTS-FE THERMAL TRANSFER ADHESIVES

11.6.1   RESULTS OF EXPERIMENTATION

Various formulations were evaluated. We learned that 60% by weight alumina powder only gave us a thermal conductivity of about 0.32 in BTU units, whereas we wanted 0.4 minimum. Adjusting the alumina up to 70% by weight gave us values above 4.0. We had also learned not to use DER 732 epoxy resin because the volume resistivity at 93°C fell below 1.0×10^{10} ohm-cm.

Using our knowledge of flexible epoxy resins and their contribution to the properties of the resultant flexible epoxy and what we had already learned about compatibility and cure characteristics with latent curatives, we arrived at a base formulation for the thermal transfer adhesive effort:

Cardolite NC-547	2 pbw
Cardolite 514	1 pbw
Epon 871	1 pbw
Latent curative	Stoichiometric amount
Alumina filler	70% by weight of formulation

We evaluated different curatives in this base formula.

11.6.2   RESULTS OF FORMULATING

Due to the fact that thermal transfer adhesives are highly loaded with alumina powder or other conductive fillers combined with the fact that high levels of curative powder is needed to formulate the 1C-RTS-FE adhesive, it was necessary to experiment with different curative powders to find ones with a favorable packing efficiency. The following table shows the formulations of the room temperature-stable, one-component, thermally conductive adhesives meeting our property objectives:

TABLE 11.7 1C-RTS-FE Thermal Transfer Adhesive Formulations

Ingredient	RTS-2000	RTS-2100	RTS-2200	RTS-2300
Cardolite NC-547 resin	12.0 pbw	13.5 pbw	13.5 pbw	11.5 pbw
Cardolite NC-514 resin	6.0 pbw	6.75 pbw	6.75 pbw	11.5 pbw
Epon 871 resin	6.0 pbw	6.75 pbw	6.75 pbw	none
Ajicure AH-122 curative	6.0 pbw	none	none	7.0 pbw
Ajicure PN-23 curative	none	3.0 pbw	none	none
Ancamine 2014FG curative	none	none	3.0 pbw	none
325 mesh alumina powder	70 pbw	70 pbw	70 pbw	70 pbw

The properties of the four successful candidates are listed in the following table:

TABLE 11.8 Properties of Four 1C-RTS-FE Thermal Transfer Adhesives

Flexipoxy name	Curing rate	Thermal conductivity, Btu/h-ft-°F	Volume resistivity, ohm-cm	Dielectric strength, V/mil	Duro-meter	Lap shear strength, psi
RTS-2000	2 h @ 100°C, 1 h @ 120°C	0.42	5.0×10^{12} at 23°C 2.3×10^{10} at 93°C	350	59 A 38 D	820
RTS-2100	2 h @ 100°C, 1 h @ 120°C	0.43	6.0×10^{13} at 23°C 9.8×10^{10} at 93°C	550	86 A 44 D	890
RTS-2200	2 h @ 100°C, 1 h @ 120°C	0.43	1.0×10^{13} at 23°C 2.0×10^{10} at 93°C	465	85 A 43 D	850
RTS-2300	2 h @ 100°C, 1 h @ 120°C	0.44	6.0×10^{12} at 23°C 3.0×10^{10} at 93°C	450	88 A 47 D	950

11.7   FORMULATION DEVELOPMENT OF 1C-RTS-FE ELECTRICALLY CONDUCTIVE ADHESIVES

11.7.1   SELECTING THE BEST FLEXIBLE EPOXY RESIN/LATENT CURATIVE BASE

The trick is to find combination of flexible epoxy resin and latent curative, which not only cures properly but has maximized liquidity uncured. In order to obtain electrical conductivity, the metallic particles must occupy about 40% of the volume of the adhesive. We started our screening with flexible epoxy resins having higher epoxide equivalent weights. Table 11.9 shows the results of the screening. Only NC-547 with AH-127 and Heloxy 505 with AH-127 offered us hope of succeeding.

TABLE 11.9   Screening Epoxy Resin/Latent Curative Combinations for Maximum Liquidity

Flexible epoxy resin	Epoxide equiv. wt.	Ajicure PN-23	Ancamine 2014 FG	Ajicure AH-127
Epon 872	700	No cure	No cure	No cure
Heloxy 84	650	Incompatible	Incompatible	No cure
Cardolite NC-547	600	2 h at 120°C*	2 h at 120°C*	2 h at 120°C
Heloxy 505	600	Incompatible	Incompatible	2 hr. at 120°C

*Friable, no strength.

11.7.2   SILVER FILLER LEVEL

Various silver-filled formulations were evaluated. We learned that 75% by weight silver powder gave us a volume resistivity $>10^{-2}$ ohm-cm, which missed the target. However, by adjusting the silver powder level up to 85% by weight, we attained values below 10^{-2} ohm-cm.

11.7.3   SUCCESSFUL FORMULATIONS

The following table shows the formulations of the one-component, room temperature-stable, electrically conductive adhesives meeting our property objectives:

TABLE 11.10   Formulations of 1C-RTS-FE Electrically Conductive Adhesives

Ingredient	Flexipoxy RTS-3000	Flexipoxy RTS-3100	Flexipoxy RTS-32002
Cardolite NC-547 epoxy resin	13.5 pbw	11.1 pbw	5.1 pbw
Cardolite NC-514 epoxy resin	None	None	5.1 pbw
Heloxy 505 epoxy resin	None	2.7 pbw	3.4 pbw
Ajicure AH-127 curative	1.5 pbw	1.2 pbw	1.4 pbw
DuPont V-9 silver flake powder	85 pbw	85 pbw	85 pbw

The properties of the three successful candidates are listed in the following table: All three were electrically conductive, flexible down to low temperatures, and had low to moderate adhesive strength.

TABLE 11.11   Properties of 1C-RTS-FE Electrically Conductive Adhesives

Property	RTS-3000	RTS-3100	RTS-3200
Volume resistivity, ohm-cm	5×10^{-3}	6×10^{-3}	1×10^{-3}
Glass transition temperature, °C	−40	−35	−25
Durometer, Shore A/Shore D	86/50	82/45	79/38
Lap shear strength, psi			
(1) With 1 h cure	160	120	140
(2) With 2 h cure	250	180	260

11.8   SUMMARY

We have successfully extended the scope of our Flexipoxy formulating capability to include one-component, room temperature-stable compounds. Three types of adhesives were successfully developed: (1) general purpose, (2) thermal transfer, and (3) electrically conductive. Twelve new Flexipoxy adhesives have been formulated and characterized.

11.9   PATENTS

11.9.1   THERMAL TRANSFER ADHESIVE PATENT

U.S. Patent 6,060,539 titled, "Room-Temperature Stable, One-Component, Thermally-Conductive, Flexible Epoxy Adhesives" was awarded to Ralph D. Hermansen and Steven E. Lau on May 9, 2000. The application for this patent was made on June 18, 1997. The original broader patent application was filed on July 19, 1995 but was segmented into different applications.

11.9.2 THE GENERAL-PURPOSE ADHESIVE PATENT

The U.S. Patent 6,723,803 titled, "Adhesive of Flexible Epoxy Resin and Latent Dihydrazide" was awarded to Ralph D. Hermansen and Steven E. Lau on April 20,2004. This was 5 years after I had fully retired. However, the application for this patent was made on August 20, 1996, when I was working at Hughes as a contract employee. The original broader patent application was filed on July 19, 1995 but was segmented into different applications. I was a Hughes employee during this period. This patent took about 9 years to be awarded. The assignee on the patent is Raytheon Corporation, which had purchased the major part of Hughes during this long period.

If you examine the patent itself, you will find one laughable error. Someone substituted the word "Theological" for the word "Rheological" in several spots. It is amazing that no one noticed this inappropriate term.

11.9.3 THE ELECTRICALLY CONDUCTIVE ADHESIVE PATENT

U.S. Patent 5,575,956 titled, "Room-Temperature Stable, One-Component, Electrically-Conductive Flexible Epoxy Adhesives" was awarded to Ralph D. Hermansen and Steven E. Lau on November 19, 1996. The application was made on July 19, 1995.

KEYWORDS

- epoxy adhesives
- 1C-RTS-FE adhesives
- Flexipoxy

PART IV
Custom-Formulated Compounds for Automotive Electronics Applications

CHAPTER 12

AIR BAG SENSOR ENCAPSULATION

CONTENTS

Although the general properties of Aerospace electronic materials and automotive electronic materials are very similar, they differ greatly in two important ways: (1) acceptable cost and (2) acceptable speed of application. Aerospace production rates are usually small, whereas automotive production rates can be immense. In Chapter 12, Delco Electronics was getting started in the air bag deployment device business, and was having trouble finding an encapsulant, which could survive the environment of the engine compartment. We helped them attain an optimized encapsulant. In Chapters 13 and 14, next generation materials were developed to allow high production rates without the need for expensive curing equipment. In Chapter 13, the focus is on conformal coatings. In Chapter 14, the focus is on solder joint lead encapsulants.

12.1   HOW I BECAME INVOLVED WITH DELCO ELECTRONICS COMPANY

General Motors Corporation bought Hughes Aircraft Company from the Howard Hughes Medical Institute in 1985. On December 31, 1985, General Motors merged Hughes Aircraft with its Delco Electronics unit to form Hughes Electronics Corporation, an independent subsidiary. This entity then consisted of: Delco Electronics Corporation and Hughes Aircraft Company. It was expected that the two electronic companies would share technology and intellectual resources. Thus, many of us in TSD were encouraged to visit Delco Electronics in Kokomo and offer our assistance on solving problems, which they might have. I found a Delco Electronics materials engineer, who welcomed my assistance and who would become my principal contact at Delco Electronics for several years. His name was Henry Sanftleben, but everyone called him Hank. Hank, in turn, worked closely with Jim Rossen, an engineer from the processing department. The three of us ended up with our names on several patents.

During my plant tours at Delco Electronics in Kokomo, I realized how different Hughes was from them. Hughes tended to design and manufacture products, which were very expensive because they were state-of-the-art and few of them would be purchased. Delco Electronics was the opposite of that. Their products were designed to be inexpensive and millions of them would be manufactured. I saw the manufacturing line for car radios. They produced 50,000 of them each day. The manufacturing process was highly automated and a minimum of people were needed to run the operation. I had to admire how brilliant their engineers were in designing and debugging such a system.

My working relationship with Hank and Jim at Delco Electronics lasted almost until I fully retired in 1998. During these latter years, huge organization changes were underway. Our CEO, Michael Armstrong, had dismembered the former HAC. He was in the process of selling off the Space and Communications Group to Boeing and rest of the company to Raytheon. Concurrently, Delco Electronics was absorbed into Delphi Automotive Systems.

12.2   INTRODUCTION TO THE AIR BAG ENCAPSULANT PROBLEM

Hank shared a problem with me. He had been evaluating commercial potting compounds for an air bag sensor application. There would be three of the sensors in the automobile. The air bag would deploy if any two sensors or more were triggered. One of the sensors was to be located in the engine compartment, where the most extreme temperatures are experienced (i.e., 105°C). Hank was having trouble finding a potting compound, which could survive this severe environment. Of the commercial products he had evaluated, he found significant differences in how they fared when he tested them. He felt that it was important to know why some compounds performed better than others in order to work toward the optimum one. However, the suppliers would not tell him anything about the composition of the submitted compounds. They told him that such information was proprietary and could not be revealed. I had previous experience in analyzing competitor's compounds, plus access to our excellent analytical laboratory and its specialists. I told Hank to send me the compounds of interest and I would see what we could learn about them.

12.3   GENERIC IDENTIFICATION OF PROPRIETARY POTTING COMPOUNDS

In 1989, Hank was seeing significant differences between candidates during his application testing, but the vendors would not tell him anything about their composition, so he was unable to learn why the more successful candidates did better, and thus be able to focus on that generic type. We offered to help in that regard, and consequently Delco Electronics funded TSD to analytically determine the generic composition of them. The most valuable analytical test was Fourier transform infrared analysis (FTIR). Phil Magallanes operated the instrument and was extremely skilled in interpreting the spectra. My knowledge of formulating allowed me to take Phil's results and propose a likely formulation. We actually formulated equivalent epoxy and polyurethane potting compounds and conducted key tests to verify that we had matched the compound.

12.4 FORMULATING FLEXIPOXY POTTING COMPOUNDS FOR THE AIR BAG SENSOR

12.4.1 CANDIDATE SCREENING CRITERIA

During our 1989 discussions, I mentioned to Hank that we had developed a series of elastomeric epoxy compounds, which we called "Flexipoxy." I suggested that we might be able to use that knowledge to formulate a successful compound for potting the air bag sensor. Hank had funded us to investigate the feasibility. The screening criteria for our effort were as follows:

TABLE 12.1   Screening Tests for Flexipoxy Potting Compound Formulation Trials

Property	Test Method	Requirement
Shore A durometer	ASTM D2240	60–90
Part A to Part B weight ratio		Near 50:50
Brookfield viscosity	Measured at 20 rpm	1500 cps. max
Thermal stability	7 day dry heat exposure at 125°C	Minimal color change
		Minimal durometer change
		Minimal weight change
Hydrolytic stability	7 days at 95% RH and 85°C	Minimal color change
		Minimal durometer change
		Minimal weight change

Note: It was intended that sand would be added to the potting compound to reduce the cost. The hydrolytic stability test was conducted on both filled and unfilled test specimens. The thermal stability test was conducted on unfilled test specimen only. The cure schedule was 1 h at 105°C.

The Flexipoxy candidates were formulated to hopefully meet the screening test requirements. The two component versions of the formulations are made by mixing the epoxy resins together as Part A and the curing agents together as Part B. In actual production, the Part A and Part B will be metered and mixed by machine. A 50:50 by volume is most desirable for machine mixing. The weight ratio is indicative of the volume ratio because the densities of the ingredients are in a close range.

12.4.2 TRIAL FLEXIPOXY ENCAPSULANTS

Many Flexipoxy trial formulations were evaluated and some lessons were learned. Epon 871 epoxy resin was unsuitable for this application because its ester groups break down during humid aging tests. Even when used at 50% of the total resin, it was still problematic. On the other hand, we learned that DER 736 epoxy resin reduces dry heat aging performance, if used at levels much above 50% of total resin. The oxypropylene units in DER 736 have poor heat stability. Heloxy WC-67, which is the diglycidyl ether of 1,4-butane diol, imparts excellent hydrolytic and thermal stability as seen in Flexipoxy 315 where it was the sole epoxy resin. The problem was excessive exothermic heat. Dytek A curative also caused formulations to be to exothermic.

12.4.3 FLEXIPOXY POTTING COMPOUND FINALISTS

Four formulations were selected for inclusion in this chapter. Their formulations are shown in Table 12.2.

TABLE 12.2 The Best Three Flexipoxy Potting Compounds per Screening Tests

Ingredient	Flexipoxy 313	Flexipoxy 315	Flexipoxy 326	Flexipoxy 339
DER 736	24.07 wt. %	None	None	None
Epon 828	None	None	None	32.50 wt. %
Heloxy WC-67	24.07 wt. %	44.27 wt.%	26.67 wt. %	8.75 wt. %
DER 732	None	None	None	8.75 wt. %
Epon 871	None	None	26.67 wt. %	None
DP-3680	51.86 wt. %	55.73 wt.%	None	42.50 wt. %
Versamid 140	None	None	26.66 wt. %	None
Castor oil	None	None	20.00 wt. %	None
ATBN	None	None	None	7.50 wt. %

The properties of the four finalists are presented in Table 12.3.

TABLE 12.3   Properties of Best Three Flexipoxy Potting Compounds

Property	Flexipoxy 313	Flexipoxy 315	Flexipoxy 326	Flexipoxy 339
Part A/Part B weight ratio	48.1/51.9	44.3/55.7	53.0/47.0	50/50
Brookfield viscosity	112 cps	100 cps	1400 cps	1260 cps
Shore A durometer	67	72	62	82
Garner color	12	18	10	
Thermal stability: 7 days at 125°C	Very good. Shore 75A, End color = 25	Very good. Shore 76A, End color = 18	Good. Shore 74A, End color = dark	See note 1
Hydrolytic stability: 7 days at 85°C 95% RH	Very good. Shore 71A	Very good. Shore 75A	Good. Shore 73A	See note 1
Concerns	None	Excessive exotherm	None	None

Note 1: Untested, yet good results expected due to presence of Epon 828 and Heloxy WC-67. Results should be better than Flexipoxy 313, which was rated very good.

12.4.4   FURTHER EVALUATION OF FLEXIPOXY 313, 326, AND 339

Flexipoxy 315 was eliminated to concern about high exotherm in a larger mass. Liquid samples and potted sensor units were prepared and shipped to Hank for further evaluation. However, Hank evaluated a commercial potting compound, which came very close to meeting his requirements and had focused his main attention on it.

12.5   IDENTIFYING A WINNER POTTING COMPOUND

12.5.1   PC-1234 POTTING COMPOUND LOOKS PROMISING

Our Flexipoxy potting compound development effort was preempted because Hank had just found a candidate that looked very promising in meeting his requirements. The superior candidate was a two component, polyurethane potting compound, which I am giving the pseudonym, PC-1234 to protect manufacturer's identity. We were asked to determine its generic composition.

12.5.2 DETERMINING THE GENERIC COMPOSITION OF PC-1234

FTIR spectroscopy revealed that the polyol component was actually a mixture of two different polyols. These polyols were separated by liquid chromatography and the separated polyols were identified as a polybutadiene polyol and a polyoxypropylene triol using quantitative FTIR procedures. In addition, the molecular weights of the two polyols were determined by size exclusion chromatography. The molecular weight of the polybutadiene polyol was about 2900 and the polyoxypropylene triol was about 624. The identification of the polyols was confirmed to be PolyBd R45HT and Pluracol TP-440 by comparison of unknowns with known polyols using size exclusion chromatography. The percentages of the two polyols in the polyol blend were established by comparing the FTIR spectra of the PC-1234 polyol component with known mixtures of PolyBd R45HT and Pluracol TP-440. So, It appeared that the PC-1234 polyol blend was likely to be 87% PolyBd R45HT and 13% Pluracol TP-440. A solids part of the PC-1234 polyol component was identified as iron oxide, a colorant and Linde molecular sieves type 4A, which is a moisture scavenger.

The isocyanate component of PC-1234 was determined to be a liquid MDI, similar to Isonate 143L. A comparison of the PC-1234 isocyanate with Isonate 143L using high-pressure liquid chromatography (HPLC) showed them to be a good match. Figure 12.1 shows the likely structure of Isonate 143L.

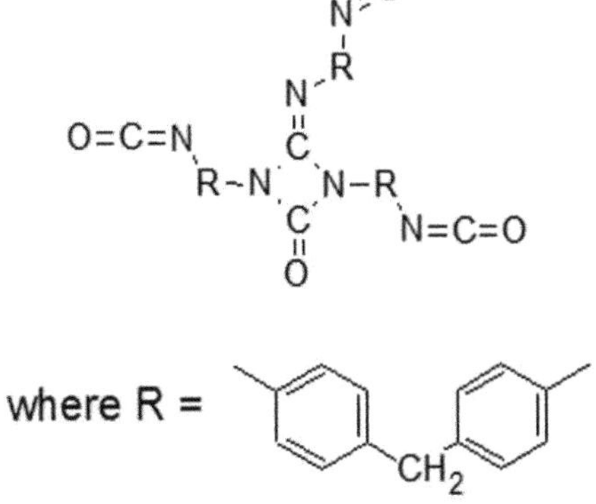

FIGURE 12.1   Likely structure for Isonate 143L Isocyanate.

12.6   DUPLICATION OF PC-1234 POTTING COMPOUND

We needed a duplicate of PC-1234, where we knew the formulation for purposes of future refinements. Although PC-1234 was the best encapsulant so far according to Delco Electronic testing, it needed to be optimized. Our duplicate encapsulant would serve as the vehicle for optimization and the lessons learned would be passed to the manufacturer of PC-1234.

12.6.1   HACTHANE 110 POTTING COMPOUND FORMULATED PER THE CHEMICAL ANALYSIS

The best way to confirm our analysis was to assemble the ingredients to make a potting compound and see if the properties are equivalent. To this end, two trial formulations were prepared using the analytical findings, one of them with a black colorant and the other without. The formulations are shown in the following table:

TABLE 12.4   Formulation of Hacthane 110 and Hacthane 110A Potting Compounds

Ingredient	Hacthane 110	Hacthane 110A
PolyBd R45HT polyol	90 pbw	90 pbw
Pluracol TP440 triol	10 pbw	10 pbw
Conap DS-1832 black colorant	None	3 pbw
DBTDL catalyst	None	1 drop
Total of polyol component	100 pbw	103 pbw
Isonate 143L diisocyanate	20 pbw	20 pbw
Total polyol plus isocyanate	120 pbw	123 pbw

Note: The cure schedule for Hacthane 100 was 2 h at 180°F, whereas the cure schedule for Hacthane 100 A was 1 h at 180°F.

12.6.2   COMPARISON OF PC-1234 VERSUS HACTHANE 110 AND PC-1234 A VERSUS HACTHANE 110A

Joan Lum in our department was asked to compare the PC-1234 with the Hughes potting compounds by running a series of tests designed to evaluate potting compounds. She did a very professional job and wrote a comprehensive report. The following table summarizes her findings.

TABLE 12.5   Property Comparison of Hacthane 110 and PC-1234 Potting Compounds

Property	PC-1234 (Brown)	Hacthane 110 (Clear)	PC-1234 A (Brown)	Hacthane 110A (Black)
Tensile strength, psi Per ASTM D638 See Note 1	178	195	260	154
Ultimate elongation, % Per ASTM D638	110	100	142	175
Tear resistance, pli Per ASTM D624	41.6	39.3	47.3	38.3
Lap shear strength, psi (PBT to PBT) (see Note 2)	200	240	210	220
Lap shear strength, psi (E-coated metal)	190	300	280	400
Durometer, Shore A Per ASTM D2240	51	53	56	52
Volume resistivity, ohm-cm Per ASTM D257	6.7×10^{14}	7.0×10^{14}	7.5×10^{14}	9.0×10^{14}
Dry heat aging (Note 3)				
a) Volume resistivity, ohm-cm	4.7×10^{15}	6.2×10^{15}	2.1×10^{15}	5.7×10^{15}
b) Shore A durometer	79	80	86	76
Humid aging (Note 4)				
a) Volume resistivity, ohm-cm	2.0×10^{13}	1.6×10^{14}	1.1×10^{14}	1.2×10^{14}
b) Shore A durometer	58	58	57	53

Note 1: The tensile specimens for PC-1234 A were provided by Delco Electronics. The specimens were tested at a crosshead speed of 0.5 inch/min.

Note 2: PBT stands for poly(butylene terephthalate)

Note 3: Samples were exposed for 7 days at 125°C. The volume resistivity and Shore A durometer were measured at the end of the exposure period.

Note 4: Samples were exposed for 7 days at 95% RH and 85°C. The volume resistivity and Shore A durometer were measured at the end of the exposure period.

12.6.4   HACTHANE 110 IS DEEMED A NEAR EQUIVALENT TO PC-1234 POTTING COMPOUND

All in all, Hacthane 110 may not be identical to PC-1234, but it is an extremely close match. The Shore A durometers are within experimental error. Tensile strengths and ultimate elongations are of similar magnitude. The tear resistance values are all very close. The lap shear strengths to PBT are very similar although there is more scatter in the lap shear strengths to E-metal. Hacthane 110 values are significantly higher. The volume resistivity results are very similar. All of these potting compounds behaved similarly in dry heat aging and in humid aging. The dry heat aging exposure had the effect of improving the volume resistivity values for the specimens markedly. This was surprising because we expected the values to fall. We also noticed that the hardness of the specimens increased rather than decreased from the dry heat exposure. Additional cross-linking may be responsible for the property changes. Oxygen may play a role in the cross-linking. The specimens had developed a hard skin, whereas the interior was still soft. The humid aging exposure had the effect of lowering the volume resistivity values of the specimens noticeably, while their durometer hardness only increased moderately.

12.7   POTTING COMPOUND OPTIMIZATION—OVERVIEW

After viewing the comparison of PC-1234 and Hacthane 110 and seeing how similar they are, Hank requested that we begin a project to optimize Hacthane 110. He intended to share our findings with PC-1234 manufacturer company so that they could use our findings to optimize their potting compound.

Hank listed the following as the most important goals for the project:

(1) Evaluate the potential for cost reduction,
(2) Improve adhesion,
(3) Improve thermal stability, and
(4) Combine findings into an optimized potting compound.

12.8 POTTING COMPOUND OPTIMIZATION—COST REDUCTION

12.8.1 PROBLEM AND APPROACH

Delco Electronics made air bag sensors for millions of cars per year in a competitive industry. Cost savings of pennies per lb can be worth hundreds of thousands of dollars.

Our approach was to find cheaper ingredients if they exist. The addition of fillers to the potting compound is one approach. Secondly, isocyanates are the most costly ingredients. A switch to a cheaper isocyanate may be feasible if properties are not sacrificed. Thirdly, finding equivalent alternatives to Isonate 143L would allow competitive bidding to be used to lower the costs.

12.8.2 COST REDUCTION VIA NON-SETTLING FILLERS

The raw material cost of Hacthane 110 was calculated to be $1.53/lb and the black pigmented version was $1.75/lb. The candidate fillers were:

Extendospheres SF-12, specific gravity = 0.75, price = $0.44/lb,

QP Corp Q-Cel, specific gravity = 0.60, price = $0.28/lb,

Z-Light G-3400, specific gravity = 0.73, price = $0.38/lb.

Note: These three fillers have a lower specific gravity than unfilled Hacthane 110, which is 0.94. Although they will not settle, they will float. They were as close to the specific gravity of Hacthane 110 as we could find. Mineral fillers, such as clay, silica, calcium carbonate, etc. have much higher specific gravities than Hacthane 110 and will settle out.

The three low-density fillers were mixed into Hacthane 110 polyol at various levels and the viscosity determined with a Brookfield RVT Viscometer using spindle #29, and 20 rpm.

TABLE 12.6   Viscosity versus Filler Content for Filled Hacthane 110 Polyol Component

Filler content, pbw	Viscosity of polyol/ SF-12 mixture, cps	Viscosity of polyol/ G3400 mixture, cps	Viscosity of polyol/ Q-Cel mixture, cps
10	10,500	11,250	14,000
20	16,500	19,000	26,100
30	25,500	28,800	>50,000
40	45,800	>50,000	>50,000
50	>50,000	>50,000	>50,000

12.8.3   TWO NEUTRAL DENSITY FILLERS

Two fillers, which looked very promising, are: (1) 60 mesh urea formalde-hyde powder, and (2) Primax R-2080, which is a powdered rubber. Both of them have densities close to the density of Hacthane 110, so the tendency to separate is minimal. Both raised the viscosity of the polyol to 10,000 cps at about 10% filler loading. The urea formaldehyde powder, being a hard plastic, raised the durometer of the cured potting compound slight-ly. Primax R-2080, being a rubber did not raise the durometer. Primax R-2080 had another benefit. It imparts a black color to the polyol. This could eliminate the need for a colorant.

12.8.4   COST REDUCTION VIA LOWER COST ISOCYANATES

Isonate 143L is the isocyanate ingredient used in Hacthane 110. It costs $1.91/lb at the time of this study. PAPI 901 is a similar, but cruder isocya-nate sometimes used to formulate polyurethane elastomers. It costs $1.27/lb. We compared the tensile properties of full and partial replacement for-mulations. Table 12.7 shows the potential cost saving by using PAPI 901 in place of Isonate 143L, but Table 12.8 shows that the cost in terms of poorer mechanical properties is not worth the cost savings.

TABLE 12.7   Cost Savings Using PAPI 901 for Isonate 143L

Formulation	Composition	$/lb	$/cu. in.
Hacthane 110	100 pbw Hacthane 110 polyol	1.53	0.0519
	21 pbw Isonate 143L		
Hacthane 110 50% 901	100 pbw Hacthane 110 polyol	1.48	0.0500
	10.5 pbw Isonate 143L		
	10.5 pbw PAPI 901		
Hacthane 110 100% 901	100 pbw Hacthane 110 polyol	1.42	0.0482
	21 pbw PAPI 901		

TABLE 12.8   Tensile Properties of Isonate 143L/PAPI 901-Modified Hacthane 110

Formulation	Tensile strength + standard deviation	Ultimate elongation + standard deviation	100% Modulus + standard deviation
Hacthane 110	385 psi (77 psi)	291% (64%)	175 psi (10 psi)
Hacthane 110 50% 901	267 psi (30 psi)	135% (23%)	222 psi (1 psi)
Hacthane 110 100% 901	212 psi (29 psi)	233% (15%)	124 psi (14 psi)

Note: The control, Hacthane 110, has higher standard deviations because only four specimens were tested versus six for the 50/50 version and five for the 100% 901 version.

12.8.5   FINDING EQUIVALENTS TO ISONATE 143L

Three isocyanates were found in our vendor survey, which claimed to be direct substitutes for Isonate 143L. Once verified, cost savings could be realized by purchasing from the lowest bidder. The following table lists the isocyanate candidates and their properties.

TABLE 12.9   Description of Test Isocyanates

Tradename and supplier	Viscosity, cps	% NCO	Equivalent weight	Evaluation purpose
Isonate 143L, Dow chemical	31	29.2	144	Control
Lupranate MM103 BASF	40	29.5	142	Equivalent to Isonate 143L
Mondur CD Mobay Chemical	40	29–30	143	Equivalent to Isonate 143L
Rubinate LF168 ICI	35	29.3	143	Equivalent to Isonate 143L

12.8.6   ESTABLISHING EQUIVALENCY

The isocyanates claiming to be equivalent to Isonate 143L were combined with either Hacthane 110 polyol or with PC-1234 polyol, cured, cut into specimens, and tested for tensile strength, ultimate elongation, 100% modulus, tear resistance, and volume resistivity. The tables below show the results.

TABLE 12.10   Test Results for the Elastomer of Hacthane 110 Polyol and Test Isocyanate

Isocyanate	Shore A durometer	Tensile strength, psi	Ultimate elongation, %	100% Modulus, psi	Tear resistance, pli	Volume resistivity, ohm-cm
Isonate 143L	60	318	104	303	37.2	5.0×10^{14}
Lupranate MM103	63	328	106	315	36.8	4.6×10^{14}
Rubinate LF168	61	307	103	298	38.0	6.0×10^{14}
Mondur CD	60	293	102	287	38.1	4.7×10^{14}

Table 12.11   Test Results for the Elastomer of PC-1234 Polyol and Test Isocyanate

Isocyanate	Shore A durometer	Tensile strength, psi	Ultimate elongation, %	100% Modulus, psi	Tear resistance, pli	Volume resistivity, ohm-cm
Isonate 143L	63	352	105	338	33.7	5.1×10^{15}
Lupranate MM103	65	331	103	325	39.0	2.4×10^{14}
Rubinate LF168	64	316	97	326	38.0	3.3×10^{15}
Mondur CD	65	348	95	356	36.7	4.3×10^{15}

The test data clearly shows that any of the four different isocyanates could be interchanged for any of the others and the properties would be nearly identical.

12.8.7   CONCLUSIONS FROM COST REDUCTIONS STUDIES

The conclusions from the cost reduction effort are as follows:

1. Fillers: The most attractive finding is that Primax R-2080, which is a powdered rubber, could reduce the cost as a filler, neither float

nor sink due to its similar density to Hacthane 110, and eliminate the need for the expensive black colorant. It would need to be tested to assure there are no unforeseen problems.

2. Isocyanates: All in all, the small cost savings of using a cheaper isocyanate do not seem to be worth the noticeable sacrifice in tensile properties.

3. Equivalent isocyanates: Three isocyanate products were shown to be so nearly identical to Isonate 143L that they could be used interchangeably with it and the potting compound properties would be acceptable.

12.9   POTTING COMPOUND OPTIMIZATION—BETTER ADHESION PART 1

12.9.1   THE PROBLEM AND APPROACH

The potting compound must adhere to all the different materials it contacts and maintain good adhesion for the life of the vehicle. Thermal cycles cause differential expansion and contraction to occur, which the potting compound must accommodate. Humidities between arid and tropical also affect the materials of the potting compound and adherends.

Adhesion promoters, such as silanes and titanates, were evaluated as to whether they improved adhesion of Hacthane 110 to: epoxy glass PWB's, solder resist, PVC-coated wire, PE acetate-coated wire, PBT plastic, iron oxide ceramics, and Kraton elastomer.

12.9.2   DEVELOPING BASIC ADHESION DATA

First, the unmodified potting compounds adhesion to PBT plastic was evaluated to serve as a baseline reference.

The two potting compounds had similar adhesive strengths and modes of failure with the exception of the after 1 week at 85°C, 95% RH results. Hacthane 110A performed better under this condition.

TABLE 12.12   Lap Shear Strength of Hacthane 110A and PC-1234 A to PBT Substrates

Conditioning prior to Testing	Lap shear strengths for Hacthane 110A	Lap shear strengths for PC-1234 A
1 week at ambient	159 psi (SD = 21 psi)	134 psi (SD = 22 psi)
	Mode = 90% adhesive	Mode = 90% adhesive
1 week at 125°C	187 psi (SD = 15 psi)	177 psi (SD = 10 psi)
	Mode = 100% cohesive	Mode = 100% cohesive
1 week at 85°C, 95% RH	213 psi (SD = 14 psi)	161 psi (SD = 36 psi)
	Mode = 100% cohesive	Mode = 60% adhesive

12.9.3   TESTING-MODIFIED POTTING COMPOUNDS FOR IMPROVED ADHESION

Next, the influence of different additives on adhesion was evaluated. The data are presented in the tables below.

TABLE 12.13   Lap Shear Strength of PC-1244-Modified Hacthane 110A and PC-1244-Modified PC-1234 A to PBT Substrates*

Conditioning prior to testing	Lap shear strengths for modified Hacthane 110A	Lap shear strengths for modified PC-1234 A
1 week at ambient	381 psi (SD = 27 psi)	240 psi (SD = 18 psi)
	Mode = no data	Mode = no data
1 week at 125°C	389 psi (SD = 38 psi)	Not tested
	Mode = no data	
1 week at 85°C, 95% RH	392 psi (SD = 14 psi)	Not tested
	Mode = no data	

*Modification was the addition of 10 drops of PC-1244 defoamer to 500 g of polyol

Comments:

This formulation modification of adding PC-1244 defoamer yielded very beneficial results for both potting compounds. Although we have no heat aging or humidity aging results for PC-1234 A, it is reasonable to assume improvements in both conditions due to the similarity of the potting compounds and due to the improvement seen for the 1 week at ambient aging.

TABLE 12.14 Lap Shear Strength of NC-547-Modified Hacthane 110A and NC-547-Modified PC-1234 A to PBT Substrates*

Conditioning prior to testing	Lap shear strengths for modified Hacthane 110A	Lap shear strengths for modified PC-1234 A
1 week at ambient	361 psi (SD = 22 psi)	285 psi (SD = 33 psi)
	Mode = 70% cohesive	Mode = no data
1 week at 125°C	653 psi (SD = 57 psi)	Not tested
	Mode = 90% cohesive	
1 week at 85°C, 95% RH	467 psi (SD = 29 psi)	Not tested
	Mode = 70% cohesive	

*Modification consisted of replacing 10% of the polyol component with Cardolite NC-547 epoxy resin.

Comments:

This formulation modification of replacing 10% of the polyol component with Cardolite NC-547 epoxy resin yielded very beneficial results for both potting compounds. The ambient-aged-modified Hacthane 110A was an improvement, but with heat aging the effect was nearly doubled. Again, although we have no heat aging or humidity aging results for PC-1234 A, it is reasonable to assume improvements in both conditions due to the similarity of the potting compounds and due to the improvement seen for the 1 week at ambient aging.

TABLE 12.15 Lap Shear Strength of Epoxy-Silane-Modified Hacthane 110A and Amino-Silane-Modified Hacthane 110 to PBT Substrates*

Conditioning prior to testing	Lap shear strengths for A-186 silane-modified Hacthane 110A	Strengths for A-1100 silane-modified Hacthane 110A
1 week at ambient	292 psi (SD = 20 psi)	246 psi (SD = 21 psi)
	Mode = 90% adhesive	Mode = 100% adhesive
1 week at 125°C	364 psi (SD = 27 psi)	311 psi (SD = 10 psi)
	Mode = 100% cohesive	Mode = 100% cohesive
1 week at 85°C, 95% RH	372 psi (SD = 11 psi)	342 psi (SD = 13 psi)
	Mode = 100% cohesive	Mode = 95 % cohesive

* Modification was the addition of 0.5 g Union Carbide A-186 epoxy functional silane to 100 g of polyol or the addition of 0.5 g Union Carbide A-1100 amino functional silane to 100 g of polyol.

Comments:

Both A-186 and A-1100 silanes improved the lap shear strength of Hacthane 110A to PBT plastic. The epoxy functional silane, A186, yielded better adhesion than the amino-silane A-1100.

12.9.4 THE ADHESION OF OPTIMIZED POTTING COMPOUNDS

Optimized versions of both Hacthane 110 and PC-1234 were formulated using the results of the preliminary adhesion. The Irganox 1076 was added from information gained in the concurrent thermal stability study. Table 12.16 shows the formulations and lap shear strengths attained using the three different materials of major interest as adherends.

TABLE 12.16   Optimized Hacthane 100 and PC-1234 Potting Compound Formulations

Formulations	Adherend material	Lap shear strength, psi	Standard deviation, psi
HACTHANE 110A optimized			
PolyBd R45HT 80 pbw			
Pluracol TP-440 10 pbw	PBT	243.6	43.9
Cardolite NC-547 10 pbw			
Irganox 1076 1 pbw	G-10	287.7	28.3
PC-1244 2 drops			
Isonate 143L 20 pbw	E-metal	302.2	14.1
PC-1234 A optimized			
PC-1234 polyol 90 pbw	PBT	277.7	26.0
Cardolite NC-547 10 pbw			
Irganox 1076 1 pbw	G-10	366.7	12.4
PC-1244 2 drops			
PC-1234 isocyanate. 21 pbw	E-metal	373.8	25.7

12.9.5 CONCLUSIONS

Major improvements in the adhesion of both potting compounds were attained by adding Cardolite NC-547 and PC-1244 defoamer to the formulations.

12.10 POTTING COMPOUND OPTIMIZATION—BETTER ADHESION PART 2

12.10.1 A BIGGER, MORE COMPREHENSIVE, PARAMETRIC TEST PROGRAM

In the year following the "Better Adhesion Part 1 Study," Hank wanted to expand the new test program to more than being simply an adhesion study. Other properties included were durometer and volume resistivity. The base formulations of Hacthane 110A and PC-1234 A were modified with 0, 5, 10, and 15% NC-547 to establish the optimal amount. Three different lap shear test substrates examined were: PBT to PBT, aluminum to aluminum, and E-metal to E-metal. The different conditioning prior to testing included: ambient, 2 or 4 weeks at 125°C, and 2 or 4 weeks at 85°C/95% RH. The following table shows the overall experimental plan.

TABLE 12.17 Matrix of Total Variables in Parametric Adhesion Study

Base formulas	% NC-547	Adherends	Conditions	Duration
PC-1234 A	0, 5, 10, 15	E-metal, aluminum, PBT plastic	Ambient, 85°C/95% RH, 125°C	Initial, 2 weeks, 4 weeks
Hacthane 110A	0, 5, 10, 15	E-metal, aluminum, PBT plastic	Ambient, 85°C/95% RH, 125°C	Initial, 2 weeks, 4 weeks
Hacthane 110A with antioxidant and defoamer	0, 5, 10, 15	E-Metal, aluminum, PBT plastic	Ambient, 85°C/95% RH, 125°C	Initial, 2 weeks, 4 weeks

12.10.1.1   VARIABLE COMBINATIONS ACTUALLY TESTED

An experimental design was used to reduce the amount of testing from this huge parametric study. Even so, 720 tests were conducted in this extensive evaluation. Table 12.18 below shows the actual variable combinations, which were selected for testing.

TABLE 12.18   Actual Variable Combinations Selected for Testing

Base formula	NC-547 added	Substrate	Condition	Duration
PC-1234 A	0, 5, 10, and 15%	PBT	Ambient	Initial
PC-1234 A	0, 5, 10, and 15%	Aluminum	125°C	2 weeks
PC-1234 A	0, 5, 10, and 15%	E-metal	85°C/ 95% RH	4 weeks
Hacthane 110 A	0, 5, 10, and 15%	PBT	125°C	4 weeks
Hacthane 110 A	0, 5, 10, and 15%	Aluminum	85°C/ 95% RH	Initial
Hacthane 110 A	0, 5, 10, and 15%	E-metal	Ambient	2 weeks
Modified Hacthane 110A	0, 5, 10, and 15%	PBT	85°C/ 95% RH	2 weeks
Modified Hacthane 110A	0, 5, 10, and 15%	Aluminum	125°C	4 weeks
Modified Hacthane 110A	0, 5, 10, and 15%	E-metal	Ambient	Initial

12.10.2   REPORTING PARAMETRIC ADHESION STUDY TEST RESULTS

Instead of reporting all of these copious data to the reader, I have summarized the results in the paragraphs below as applicable.

12.10.3   PARAMETRIC ADHESION STUDY—VOLUME RESISTIVITY RESULTS

For both Hacthane110 and PC-1234 and all the formula variations of them, the volume resistivity values were excellent. All of the values measured would lie between 1×10^{14} and 1×10^{17} ohm-cm. As we observed in the 1990 optimization effort, dry heat aging tends to increase volume resistivity slightly, whereas humid/heat aging tends to lower volume resistivity slightly. Formulation wise, the addition of Cardolite NC-547 to each of the base formulas at 0, 5, 10, and 15% seemed to be unrelated to the volume resistivities of the samples.

12.10.4 PARAMETRIC ADHESION STUDY—SHORE A DUROMETER RESULTS

I felt that there was sufficient interpretive value in the durometer data to present nearly all of it. The table below shows the Shore A durometer results of the various formulations aging at ambient conditions.

TABLE 12.19   Shore A Durometer of Hacthane 110 and PC-1234 versus NC-547 Content versus Ambient Aging

Test potting compound	Shore A initial	Shore A at 7 days	Shore A at 14 days	Shore A at 28 days
Hacthane 110 + no NC-547	53	74	75	75
PC-1234 + no NC-547	55	64	71	74
Hacthane 110 + 5% NC-547	58	75	75	76
PC-1234 + 5% NC-547	59	73	72	73
Hacthane 110 + 10% NC-547	62	75	75	76
PC-1234 + 10% NC-547	63	75	74	74
Hacthane 110 + 15% NC-547	64	75	76	76
PC-1234 + 15% NC-547	65	75	75	74

Notice that the increasing addition of Cardolite NC-547 tends to raise the initial durometer of both Hacthane 110 and PC-1234. However after a week, all of the trials had a nearly identical durometer of 75 and that it did not change thereafter.

The durometer results after dry heat aging of the same formulations were initially identical, but rose dramatically thereafter.

We see some interesting trends in these data that we did not see in the ambient aging data. First of all, the durometer rises with increasing NC-547 content for all the four durations of aging. This might be explained by the higher temperature exposure, where new cross-linking reactions are possible. Secondly, comparing the regular with the modified Hacthane data, we see that the antioxidant decreases the durometer rise for about 1 week or so and then the effect is gone with longer heat durations.

Data on regular PC-1234 during dry heat aging was obtained but suffice it to say that it is the same as for the regular Hacthane 110 presented above. PC-1234 with antioxidant was neither prepared nor tested.

TABLE 12.20   Shore A Durometer of Hacthane 110 versus Stabilized Hacthane 110 versus NC-547 Content versus Dry Heat Aging

Test potting compound	Shore A initial	Shore A at 7 days	Shore A at 14 days	Shore A at 28 days
Regular Hacthane 110 + no NC-547	53	84	92	92
Modified Hacthane 110 + no NC-547	54	80	82	91
Regular Hacthane 110 + 5% NC-547	58	87	94	93
Modified Hacthane 110 + 5% NC-547	58	82	87	95
Regular Hacthane 110 + 10% NC-547	62	91	95	96
Modified Hacthane 110 + 10% NC-547	63	88	94	97
Regular Hacthane 110 + 15% NC-547	64	93	95	97
Modified Hacthane 110 + 15% NC-547	64	88	95	98

Note: Modified Hacthane 110 contains PC-1244 defoamer and Irganox 1076 antioxidant.

Table 12.21 shows how durometer changes with exposure to humid heat aging.

TABLE 12.21   Shore A Durometer of Hacthane 110 versus Stabilized Hacthane 110 versus NC-547 Content versus Humid Heat Aging

Test potting compound	Shore A initial	Shore A at 7 days	Shore A at 14 days	Shore A at 28 days
Regular Hacthane 110 + no NC-547	53	74	76	77
Modified Hacthane 110 + no NC-547	54	75	75	75
Regular Hacthane 110 + 5% NC-547	58	77	79	79
Modified Hacthane 110 + 5% NC-547	58	76	76	78
Regular Hacthane 110 + 10% NC-547	62	79	82	83
Modified Hacthane 110 + 10% NC-547	63	78	80	81
Regular Hacthane 110 + 15% NC-547	64	78	82	82
Modified Hacthane 110 + 15% NC-547	64	80	81	82

Note: Modified Hacthane 110 contains PC-1244 defoamer and Irganox 1076 antioxidant.

Comments:

Again, we see some interesting trends in these data that we did not see in the ambient aging data. First of all, the durometer rises with increasing NC-547 content for all the four durations of aging. It is not as pronounced as with the dry heat aging, which was a higher temperature exposure. Moreover, the humidity would preferentially use up free isocyanate to form urea linkages. Secondly, comparing the regular with the modified Hacthane data, we see that the antioxidant decreases the durometer rise for the entire 28-day duration. Again, the exposure is at a significantly lower temperature. Finally, the same trends were seen in the PC-1234 humid aging data, which was omitted for brevity here.

12.10.5   PARAMETRIC ADHESION STUDY—LAP SHEAR STRENGTH RESULTS

The following three tables show the lap shear strength data for the sets selected.

TABLE 12.22   Lap Shear Strength of NC-547-Modified PC-1234 A

Adherend substrate	Exposure	LS strength 0% NC-547	LS strength 5% NC-547	LS strength 10% NC-547	LS strength 15% NC-547
PBT	Ambient	156 psi	272 psi	298 psi	299 psi
	Initial	SD = 22 psi	SD = 32 psi	SD = 21 psi	SD = 15 psi
Aluminum	125°C	203 psi	293 psi	261 psi	265 psi
	for 2 weeks	SD = 38 psi	SD = 32 psi	SD = 31 psi	SD = 21 psi
E-metal	85°C/95% RH	299 psi	294 psi	331 psi	335 psi
	Initial	SD = 20 psi	SD = 27 psi	SD = 12 psi	SD = 28 psi

TABLE 12.23   Lap Shear Strength of NC-547-Modified Hacthane 110A

Adherend substrate	Exposure	LS strength 0% NC-547	LS strength 5% NC-547	LS strength 10% NC-547	LS strength 15% NC-547
E-metal	Ambient for 2 weeks	375 psi	376 psi	402 psi	377 psi
		SD = 19 psi	SD = 50 psi	SD = 13 psi	SD = 22 psi
PBT	125°C for 4 weeks	230 psi	267 psi	359 psi	374 psi
		SD = 30 psi	SD = 17 psi	SD = 26 psi	SD = 43 psi
Aluminum	85°C/95% RH	288 psi	303 psi	304 psi	318 psi
	Initial	SD = 12 psi	SD = 33 psi	SD = 26 psi	SD = 29 psi

TABLE 12.24 Lap Shear Strength of NC-547-Modified Hacthane 110A/PC1244/IRG1076

Adherend substrate	Exposure	LS strength 0% NC-547	LS strength 5% NC-547	LS strength 10% NC-547	LS strength 15% NC-547
E-Metal	Ambient	196 psi	218 psi	229 psi	278 psi
	Initial	SD = 13 psi	SD = 26 psi	SD = 24 psi	SD = 25 psi
Aluminum	125°C for 4 weeks	196 psi	225 psi	221 psi	245 psi
		SD = 16 psi	SD = 26 psi	SD = 8 psi	SD = 29 psi
PBT	85°C/95% RH	288 psi	303 psi	304 psi	318 psi
	for 2 weeks	SD = 12 psi	SD = 33 psi	SD = 26 psi	SD = 29 psi

Comments:

In general, all of the lap shear strength data falls within a narrow range, from 156 to 402. The three base formulations seem very similar to one another. Increasing amount of NC-547 seemed to improve lap shear strength overall. Adhesion to PBT was clearly improved by NC-547.

12.10 POTTING COMPOUND OPTIMIZATION—BETTER THERMAL STABILITY PART 1

12.10.1 THE PROBLEM

Hacthane 110 is predominately a polybutadiene polyurethane. Having no ester and only a few ether groups in the compound enhances its hydrophobicity and its dielectric properties. However, it obtains its elastomeric nature from frequent carbon to carbon double bonds in the polybutadiene segment and those double bonds are subject to oxidation. What we observe with Hacthane 110 and PC-1234 elastomers during dry heat aging at 125°C is a hard skin progressively developing while the interior of the specimens remains soft. This hardening process causes the potting compound to lose its needed elastomeric properties.

12.10.2 THE APPROACH

The action is to find suitable antioxidants and UV stabilizers for polyurethane compounds and compare their effectiveness in improving dry heat aging and humidity heat aging performance. We discussed the topic of thermal stability with Atochem representatives, makers of PolyBd R45HT polyol, and they suggested we evaluate the following antioxidants: Irganox 1076, Irganox 1010, and triphenylphosphine.

12.10.3   STABILIZER EFFECTIVENESS—TEST RESULTS

Measuring the durometer of the cured potting compounds was deemed an effective way to monitor the retention of flexibility under aging conditions. First let us examine the most severe aging condition, dry heat aging. The following table shows the results of Hacthane 110A and PC-1234 A containing different antioxidants employed at 1% level to the polyol component.

TABLE 12.25   Stabilization of Hacthane 110A and PC-1234 A to Dry Heat Aging

Potting compound	Antioxidant	Durometer, initial	Durometer, 2 days at 125°C	Durometer, 7 days at 125°C
Hacthane 110A	None	53 A	78 A	88 A
Hacthane 110A	1010	53 A	66 A	68 A
Hacthane 110A	1076	53 A	77 A	78 A
Hacthane 110A	TPP	53 A	78 A	88 A
PC-1234 A	None	55 A	67 A	88 A
PC-1234 A	1010	55 A	56 A	59 A
PC-1234 A	1076	55 A	61 A	61 A
PC-1234 A	TPP	55 A	77 A	88 A

Comments:

For both Hacthane 110A and PC-1234 A, Irganox 1010 was most effective in stabilizing the durometer change, Irganox 1076 was next most effective, and triphenylphosphine was ineffective. PC-1234 A seemed to be better stabilized by the antioxidants than was Hacthane 110A. We suspected that the difference might be due to more free isocyanate in Hacthane 110A than in PC-1234 A. The former has an NCO/OH ratio of 1.10 versus 0.95 for the latter potting compound. Excess isocyanate makes cross-linking reactions possible, which raise the durometer. At the time of this study (1990), the cost of Irganox 1076 was $4.51/lb and the cost of Irganox was $ 6.99/lb for quantities over 550 lbs in 1990. Employed at 1% level, they raise the raw material cost four to seven cents per pound.

12.10.4 DETERMINING THE INFLUENCE OF NCO/OH RATIO ON ROOM TEMPERATURE AMBIENT AGING

We believed that it was necessary to determine the influence of NCO/OH ratio on initial durometers and on subsequent aging at room temperature in order to fully understand what was going on. Therefore, both Hacthane 110 A and PC-1234 were evaluated at NCO/OH ratios of 0.8, 0.9, 1.0, 1.1, and 1.2. No antioxidant was added to either compound for this study. Ideally, the initial durometer versus NCO/OH ratio curve should have a maximum, that is, where the mole ratio of reactants is equal. The 7-day durometer versus NCO/OH ratio curve should show higher durometers especially as more isocyanate is available for cross-linking reactions. The following table shows the results.

TABLE 12.26 Durometer Change during Room Temperature Aging versus NCO/OH Ratio

Test compound, days at room temperature	NCO/OH = 0.8	NCO/OH = 0.9	NCO/OH = 1.0	NCO/OH = 1.1	NCO/OH = 1.2
Hacthane 110 A Initial durometer Shore A	44	50	57	59	59
Hacthane 110 A 7-day durometer Shore A	52	59	70	75	78
PC-1234 A Initial durometer Shore A	40	55	63	65	64
PC-1234 A 7-day durometer Shore A	40	56	68	75	77

Comments:

The shape of the initial durometer versus NCO/OH ratio curves is very similar for both test compounds. The ideal NCO/OH value for Hacthane 110 A is about 1.15 and the ideal NCO/OH value for PC-1234 A is about 1.10. The initial durometer curve versus NCO/OH ratio curves are at a maximum at these ideal values and a minor mixing error in either direction still yields about the same Shore hardness. The 7-day-aged durometer versus NCO/OH ratio are also very similar for both compounds. These curves for both compounds increase as NCO/OH ratio increases. This observation is consistent with the fact that free NCO groups can undergo reaction with water being absorbed to form a urea cross-link. The more free NCO, the more cross-linking occurs.

12.10.5 DETERMINING THE INFLUENCE OF NCO/OH RATIO ON DRY HEAT THERMAL STABILITY

We also felt that it was important to understand the relationship between NCO/OH ratio and dry heat thermal stability. Therefore, both Hacthane 110 A and PC-1234 were evaluated at NCO/OH ratios of 0.8, 0.9, 1.0, 1.1, and 1.2. No antioxidant was added to either of the compound for this study. The following table shows the results.

TABLE 12.27 Durometer Change during Dry Heat Aging versus NCO/OH Ratio

Test compound, days at 125°C	NCO/OH = 0.8	NCO/OH = 0.9	NCO/OH = 1.0	NCO/OH = 1.1	NCO/OH = 1.2
Hacthane 110 A Initial durometer Shore A	44	50	57	59	59
Hacthane 110 A 7-day durometer Shore A	83	87	89	92	94
PC-1234 A Initial durometer Shore A	40	55	63	65	64
PC-1234 A 7-day durometer Shore A	84	86	88	90	91

Comments:

If we compare Tables 12.25 and 12.26, we see that the 7-day durometers are much higher due to the dry heat conditioning. In fact, the potting compounds have lost most or all of their elastomeric nature. As anticipated, the effect is worse at higher NCO/OH ratios. The fact is that free NCO groups can undergo allophanate and biuret cross-linking reactions at elevated temperatures. The more free NCO, the more cross-linking occurs. Concurrently, cross-linking via carbon–carbon double bonds in the polybutadiene chains is occurring.

12.10.6   CONCLUSIONS

1. Dry heat aging at 125°C is the most severe of the aging conditions and causes the most changes. The elastomer hardens up and loses its rubber-like properties more and more over time. By slicing up the specimens, we learned that the hardening was most severe at the exterior surface of the specimens (the air side) and diminishes inward. In other words, the outside of the specimen was hard, whereas the interior remained soft. Measuring durometer is an inexpensive and fast method of quantifying these phenomena.
2. Stabilizers tended to reduce the degree of hardening up in dry heat aging. Irganox 101 was most effective, then Irganox 1076, whereas TTP was ineffective.
3. The Hacthane 110 and PC-1234 potting compound were evaluated for NCO/OH ratio effects by measuring the durometer initially and after room temperature aging. The two compounds behave similarly, but not identically. Excess isocyanate results in higher durometers after aging as expected.
4. When the NCO/OH variation study was performed using dry heat aging, the two compounds behaved similarly and excess NCO lead to higher durometers after aging.

12.11   POTTING COMPOUND OPTIMIZATION—BETTER THERMAL STABILITY PART 2

12.11.1   PARAMETRIC TEST PLAN FOR THERMAL STABILITY STUDY PART 2

In the previous thermal stability study, Irganox 1076 at 1% appeared to be the most effective. However, the dry heat exposure phenomenon was only examined for 7 days. So in the follow-on study (Part 2), Hank wanted us to conduct a more expansive study with more antioxidants and for longer times. The new parametric thermal stability study is summarized in Table 12.28.

TABLE 12.28 The Parameters of the New Expanded Thermal Stability Study

Test antioxidant	Percentage in Hacthane 110	Ambient exposure	125°C dry exposure	85°C/95% RH exposure
Cyanox blend See Note 1	0.0175, 0.350, and 0.525	1, 2, 3, and 4 weeks See Note 2	1, 2, 3, and 4 weeks See Note 2	1, 2, 3, and 4 weeks See Note 2
Tenox TBHQ	0,02, 0.06, and 0.10	1, 2, 3, and 4 weeks See Note 2	1, 2, 3, and 4 weeks See Note 2	1, 2, 3, and 4 weeks See Note 2
Irganox 1010	0.50, 1.00, and 1.50	1, 2, 3, and 4 weeks See Note 2	1, 2, 3, and 4 weeks See Note 2	1, 2, 3, and 4 weeks See Note 2
Irganox 1076	0.50, 1.00, and 1.50	1, 2, 3, and 4 weeks See Note 2	1, 2, 3, and 4 weeks See Note 2	1, 2, 3, and 4 weeks See Note 2

Note 1: Cyanox blend is one part Cyanox 2777 plus 2.5 parts Cyanox LTDP

Note 2: Weekly testing consists of measuring durometer, tensile properties, and volume resistivity

12.11.2 GENERAL COMMENTS ON THE PRESENTATION OF THERMAL STABILITY RESULTS

A huge amount of test data was developed. Instead of presenting all of the details here, I will generalize as to what the results told us.

12.11.3 THERMAL STABILITY RESULTS—DUROMETER OF MODIFIED HACTHANE 110 VERSUS WEEKS OF AMBIENT AGING

The following table shows the change in durometer versus ambient aging. The unmodified Hacthane 110 aged in an acceptable manner and the stabilized versions were even more stable. The three levels for each stabilizer are presented as a group range rather than reporting them individually.

TABLE 12.29 Hacthane 110 and Stabilized Versions versus Shore A Durometer under Ambient Aging

Hacthane 110 plus	Initial	1 Week	2 Weeks	3 Weeks	4 Weeks
No additives	56	74	76	76	76
Irganox 1076 at three levels	55–61	65–70	65–69	64–69	64–69
Irganox 1010 at three levels	54–55	63–66	63–66	63–66	63–66
Tenox TBHQ at three levels	53–54	64	64	64	64
Cyanox blend at three levels	54–55	64	63–64	63–65	63–65

Comments:

(1) The unmodified Hacthane 110 climbed the highest, leveling out at 76 Shore A. All of the antioxidants retarded this climb to typically 65 Shore A.

(2) Irganox 1076 lowered the durometer at least by seven points. There was no trend regarding percentage used. The lowest level (0.5%) was the most effective.

(3) Irganox 1010 lowered the durometer of Hacthane 110 by at least 10 points. Levels 1.0% and 1.5% were most effective.

(4) Tenox TBHQ lowered the durometer of Hacthane 110 by 12 points. All levels produced the same durometer.

(5) Cyanox blend lowered the durometer by at least 11 points. Level used was not important.

12.11.4   THERMAL STABILITY RESULTS—TENSILE PROPERTIES OF MODIFIED HACTHANE 110 VERSUS WEEKS OF AMBIENT AGING

The tensile strength and ultimate elongation of ambient-aged Hacthane 110 variations all fell in an acceptable range. The data are reported later when comparing change in tensile properties due to dry heat aging. The ambient-aged tensile properties are presented then for purposed of comparison.

12.11.5   THERMAL STABILITY RESULTS—VOLUME RESISTIVITY OF MODIFIED HACTHANE 110 VERSUS WEEKS OF AMBIENT AGING

The volume resistivity versus ambient aging is shown in the following table. With one exception (Irganox 1076), the three levels of a stabilizer are shown.

TABLE 12.30   Hacthane 110 and Stabilized Versions—Log Volume Resistivity under Ambient Aging

Hacthane 110 plus	Initial	1 Week	2 Weeks	3 Weeks	4 Weeks
No additives	15.3*	15.2	15.0	15.0	15.1
Irganox 1076 at 1.5%	15.5	15.3	15.4	15.4	15.4
Irganox 1076 at 0.5 and 1.0%	14.9	14.9	14.9	14.9	14.9
Irganox 1010 at three levels	14.8–14.9	14.8–14.9	14.8–14.9	14.8–14.9	14.8–14.9
Tenox TBHQ at three levels	14.8–14.9	14.9	14.9	14.9	14.9
Cyanox blend at three levels	14.7–14.9	14.7–14.9	14.7–14.9	14.7–14.9	14.7–14.9

*Log volume resistivity in ohm-cm.

Comments:

First of all, the volume resistivity data are excellent for a potting compound and remain constant throughout the ambient aging duration. With one exception, the unmodified Hacthane 110 had higher volume resistivity than its modified versions. That one exception is Irganox 1076 at 1.5%. It had higher values than the control. Why this is true is unknown at this time.

   In all other cases, the addition of an antioxidant lowered the volume resistivity about half a decade.

12.11.6   THERMAL STABILITY RESULTS—DUROMETER OF MODIFIED HACTHANE 110 VERSUS WEEKS OF 85°C/95% RH AGING

The data from the durometer of Hacthane 110 and stabilized Hacthane 110 over 4 weeks of humid aging are very similar to the data from ambient aging. The stabilizers reduced the durometer by a few points.

12.11.7 THERMAL STABILITY RESULTS—TENSILE PROPERTIES OF MODIFIED HACTHANE 110 VERSUS WEEKS OF 85°C/95% RH AGING

In the following table, the tensile strength and ultimate elongation of 4-week ambient-aged specimens are compared with the tensile strength and ultimate elongation of 4-week heat and humidity-aged specimens.

TABLE 12.31   Tensile Strength and Ultimate Elongation of Hacthane 110 after 4 Weeks at 85°C/95% RH

Hacthane 110 plus	4-Week ambient tensile strength, psi	4-Week ambient ult. elong., %	4 Weeks 85°C/95% RH tensile strength, psi	4 Weeks 85°C/95% RH ult. elong., %
No additives	300	84	390	89
Irganox 1076 at 0.5%	260	70	320	85
Irganox 1076 at 1.0%	350	96	450	118
Irganox 1076 at 1.5%	400	112	500	128
Irganox 1010 at 0.5%	280	75	360	97
Irganox 1010 at 1.0%	280	69	370	90
Irganox 1010 at 1.5%	270	64	370	85
Tenox TBHQ at 0.02%	310	82	310	71
Tenox TBHQ at 0.06%	310	85	320	77
Tenox TBHQ at 0.10%	320	83	350	77
Cyanox blend at 0.0175%	290	75	330	70
Cyanox blend at 0.350%	380	105	380	85
Cyanox blend at 0.525%	300	78	360	78

Comments:

(1) The unmodified Hacthane 110 had as good or better tensile properties after humid aging as it did after ambient aging.
(2) For all of the candidates, tensile strength was higher after humid aging than it was after ambient aging. The elevated temperature (85°C) fully cured the polymer.

(3) Irganox 1076 and Irganox 1010 caused an improvement in ultimate elongations due to heat/humidity aging. With the other stabilizers, there was a tendency to lose ultimate elongation. Ultimate elongation may be more important than tensile strength for a potting compound. It must be able to stretch to accommodate stresses due to thermal expansion.

12.11.8 THERMAL STABILITY RESULTS—VOLUME RESISTIVITY OF MODIFIED HACTHANE 110 VERSUS WEEKS OF 85°C/95% RH AGING

The following table shows the log volume resistivity versus weeks of humid aging. The three levels of a stabilizer are shown as a range rather than individually with the exception of Irganox 1076 at 1.5% level.

TABLE 12.32 Hacthane 110 and Stabilized Versions—Log Volume Resistivity under 85°C/95% RH Aging

Hacthane 110 plus	Initial	1 Week	2 Weeks	3 Weeks	4 Weeks
No additives	15.3	15.1	15.2	15.0	15.0
Irganox 1076 at 1.5%	15.5	15.6	15.5	15.3	15.2
Irganox 1076 at 0.5 and 1.0%	14.8	14.9	14.8–14.9	14.7–14.8	14.7
Irganox 1010 at three levels	14.8–14.9	14.8–14.9	14.8–14.9	14.7	14.7
Tenox TBHQ at three levels	14.8–15.0	14.8–15.0	14.7–15.0	14.7–14.9	14.6–14.8
Cyanox blend at three levels	14.8–14.9	14.6–14.9	14.4–14.6	14.3–14.5	14.2–14.5

Comments:

All of the values are quite acceptable for a potting compound application. There was a general trend for all of the candidates to lose some resistivity due to humid aging. All of the stabilized candidates had slightly lower volume resistivity than the unstabilized version. The one exception was Irganox 1076 at 1.5% level. It had higher values than the control. The same candidate fared better than the control during the ambient aging evaluation.

12.11.9   THERMAL STABILITY RESULTS—DUROMETER OF MODIFIED HACTHANE 110 VERSUS WEEKS OF 125°C AGING

The following table shows the durometer of each candidate at each week of aging at 125°C.

TABLE 12.33   Hacthane 110 and Stabilized Versions versus Shore A Durometer under 125°C Aging

Hacthane 110 plus	Initial	1 Week	2 Weeks	3 Weeks	4 Weeks
No additives	58	85	93	93	93
Irganox 1076 at 0.5%	55	72	74	80	93
Irganox 1076 at 1.0%	60	71	74	78	88
Irganox 1076 at 1.5%	58	70	71	74	81
Irganox 1010 at 0.5%	56	70	72	77	84
Irganox 1010 at 1.0%	55	69	69	72	77
Irganox 1010 at 1.5%	54	67	68	71	74
Tenox TBHQ at 0.02%	53	72	85	89	90
Tenox TBHQ at 0.06%	53	69	81	87	90
Tenox TBHQ at 0.10%	53	68	76	83	90
Cyanox blend at 0.0175%	58	82	88	90	92
Cyanox blend at 0.350%	53	68	79	89	91
Cyanox blend at 0.525%	52	68	70	75	90

Comments:

(1) The unmodified Hacthane 110 climbed the highest, leveling out at 93 Shore A after 2 weeks. All of the antioxidants slowed down the age-hardening process, but did not stop it.

(2) Cyanox blend was the least effective of the antioxidants.

(3) Tenox TBHQ at all three levels was ineffective at 4 weeks. The data suggest that higher levels than 0.10% might improve results.

(4) Irganox 1076 was increasingly effective with increasing percentage used. The highest level (1.5%) was the most effective.

(5) Irganox 1010 was the most effective antioxidant of all. It was increasingly effective with increasing percentage used. The highest level (1.5%) of Irganox 1010 was the most effective of all antioxidants evaluated.

12.11.10 THERMAL STABILITY RESULTS—TENSILE PROPERTIES OF MODIFIED HACTHANE 110 VERSUS WEEKS OF 125°C AGING

The following table compares the tensile strength and ultimate elongation of each candidate at the end of either 4 weeks at ambient aging or 4 weeks of 125°C aging. The effect of dry heat aging on tensile properties is severe and the difference is best illustrated by comparing it with the ambient aging results.

TABLE 12.34   Tensile Strength and Ultimate Elongation of Hacthane 110 after 4 Weeks at 125°C

Hacthane 110 plus	4-Week ambient tensile strength, psi	4-Week ambient ult. elong., %	4-Weeks 125°C tensile strength, psi	4-Weeks 125°C ult. elong., %
No additives	300	84	850	5
Irganox 1076 at 0.5%	260	70	640	5
Irganox 1076 at 1.0%	350	96	770	12
Irganox 1076 at 1.5%	400	112	630	27
Irganox 1010 at 0.5%	280	75	670	4
Irganox 1010 at 1.0%	280	69	710	22
Irganox 1010 at 1.5%	270	64	470	31
Tenox TBHQ at 0.02%	310	82	480	8
Tenox TBHQ at 0.06%	310	85	410	4
Tenox TBHQ at 0.10%	320	83	640	5
Cyanox blend at 0.0175%	290	75	840	6
Cyanox blend at 0.350%	380	105	610	5
Cyanox blend at 0.525%	300	78	490	5

Comments:

(1) One glance at these data and we see tensile strengths are markedly increased by dry heat aging, whereas ultimate elongations are vastly reduced by it. We expect a potting compound to retain its initial properties for as long as possible. The potting compound should be soft, low modulus, and able to undergo stretching so as to not break electronics or develop tears and gaps in itself.

(2) The Tenox and Cyanox blend failed most dramatically in this regard and are eliminated as candidates.
(3) Irganox 1076 and Irganox 1010 both show a correlation between
level of antioxidant used and ultimate elongation. The highest heat-
aged ultimate elongations were obtained. The heat-aged tensile
strength data does not show the same nice correlation. Our focus
here is on having decent ultimate elongation.
(4) Irganox 1010 at 1.5% level was the most stable formulation of
them all.

12.11.11   THERMAL STABILITY RESULTS—VOLUME RESISTIVITY
OF MODIFIED HACTHANE 110 VERSUS WEEKS OF 125°C AGING

The following table shows the log volume resistivity data for the three
levels of Irganox 1076 and Irganox 1010 used to stabilize Hacthane 110
during 4-week exposure at 125°C.

TABLE 12.35   Hacthane 110 and Stabilized Versions—Log Volume Resistivity under
125°C Aging

Hacthane 110 plus	Initial	1 Week	2 Weeks	3 Weeks	4 Weeks
No additives	15.3	15.5	15.5	15.7	15.8
Irganox 1076 at 0.5%	14.8	15.4	15.4	15.6	15.8
Irganox 1076 at 1.0%	14.9	15.5	15.5	15.7	15.7
Irganox 1076 at 1.5%	15.5	15.8	15.8	15.7	15.6
Irganox 1010 at 0.5%	14.8	15.3	15.5	15.5	15.6
Irganox 1010 at 1.0%	14.9	15.5	15.6	15.6	15.6
Irganox 1010 at 1.5%	14.9	15.6	15.6	15.7	15.6

Comments:

The dry heat exposure causes the volume resistivity to gradually increase
in all cases. Remember that Tenox and Cyanox antioxidants had been already eliminated from further consideration because they allowed tensile
elongation to degrade too far.

12.11.12   THERMAL STABILITY STUDY CONCLUSIONS

(1) Irganox 1010 was clearly the most effective antioxidant, with Irganox 1076 a close second.
(2) Tensile properties were most dramatically affected of all the properties examined. The changes under heat/humidity aging were minor, but under dry heat aging were severe. The reduction in ultimate elongation was severe and the focus of our attention.
(3) Irganox 1010 should be used at 1.0% minimum to stabilize polyurethane compounds based upon PolyBd polyols.

12.12   POTTING COMPOUND OPTIMIZATION—COMBINING THE IMPROVEMENTS INTO A SINGLE FORMULATION

12.12.1   APPROACH

Significant progress was made in understanding and improving the thermal stability of PC-1234 and Hacthane 110 potting compounds. Both compounds harden in a dry heat environment from the outside inwards. This behavior seems to be unique to polybutadiene polyol-based polyurethanes. It is associated with the carbon to carbon double bonds in the polybutadiene chains. Antioxidants retard this behavior significantly. The antioxidant Irganox 1010 was the most effective one found in this effort.

The other area of significant progress was in adhesion enhancement. We noticed that the adhesion to PBT improved by being subjected to dry heat or humid heat aging. The mode changed from adhesive failure to cohesive failure, which is desirable. The strengths improve slightly too. The additives Cardolite NC-547, PC-1244 defoamer, and epoxy functional silane each independently nearly doubled the lap shear strengths.

12.12.2   HACTHANE 115—OUR OPTIMIZED FORMULATION

Hacthane 115 is our designation for a formulation, which incorporates the best features from the studies into a single compound. The formulation for Hacthane 115 is shown in Table 12.36.

TABLE 12.36   Formulation of Hacthane 115 Potting Compound

Ingredient	Source	Weight
PolyBd R45HT polyol	Atochem	79.05
Pluracol TP440 triol	BASF	10.00
Cardolite NC-547	Cardolite	7.50
H615 black colorant	Ciba-Geigy	1.50
Irganox 1010 antioxidant	Ciba-Geigy	1.00
PC-1244 defoamer	Monsanto	0.05
Total of polyol component		100.00
Isonate 143L diisocyanate	Dow Chemical	20.00
Total polyol plus isocyanate		120.00

Property improvements of Hacthane 115 over Hacthane 110 were as follows:

1. The addition of Cardolite NC-547 epoxy resin to the Hacthane 110 formulation improved the lap shear strength to PBT substrates significantly and improved the lap shear strength to E-metal or aluminum slightly. Cardolite NC-547 contains both epoxy reactive groups and hydroxyl reactive groups. The latter allow the molecule to become part of the polyurethane structure.
2. Four different antioxidants were evaluated at three different concentrations and the best of them was Irganox 1010 at 1.0–1.5%.
3. Three products were found to be equivalent to the Isonate 143L diisocyanate. They are: Lupranate MM103 from BASF, Mondur CD from Mobay Chemical, and Rubinate LF168 from ICI. Tests included durometer, tensile strength, 100% modulus, and ultimate elongation, tear resistance, and volume resistivity.

12.13   CHARACTERIZATION OF HACTHANE 115 POTTING COMPOUND

Hacthane 115 was thoroughly tested to give us assurance that the combination of ingredients provided the originally sought-after properties. Table 12.37 shows the processing properties of Hacthane 115 and Table 12.38 shows the properties of the cured compound.

TABLE 12.37   Processing Properties of Hacthane 115 Potting Compound

Property	Test data
Mix ratio (by weight)	Hacthane 115 Part A 100
	Hacthane 115 Part B 20
Brookfield viscosity, at 75°F	Hacthane 115 Part A 7000–8000 cp.
	Hacthane 115 Part B 30–35 cp.
	Mixture 5000–7000 cp.
Work-life, at 75°F	90 min
Cure schedule	1 h at 180°F

TABLE 12.38   Properties of Cured Hacthane 115 Potting Compound

Property	Test method	Test data
Density	ASTM D792	0.95 g/cc
Shore A durometer	ASTM D 2240	60
Volume resistivity	ASTM D257	$1–10 \times 10^{15}$ ohm-cm
Tensile strength	ASTM D412	322 psi
Ultimate elongation	ASTM D412	108%
100% Modulus	ASTM D412	295 psi
Tear resistance	ASTM D624	35 pli
Lap shear strength	ASTM D1002	
Aluminum to aluminum		200–300 psi
E-metal to E-metal		250–300 psi
PBT to PBT *		250–300 psi
Thermal stability	Aged 4 weeks at 250°F (125°C)	
	Volume resistivity change	None
	Durometer change	Hardened to 92–98
Hydrolytic stability	Aged 4 weeks at 180°F/95% RH	
	Volume resistivity change	No change
	Durometer change	Hardened to 77–83

* PBT stands for poly(butylene terephthalate)

12.14   POTTING COMPOUND PATENTS

12.14.1   HUGHES AIRCRAFT COMPANY—U.S. PATENT 5,457,165

The development work on Flexipoxy encapsulants was of no further interest to Delco Electronics once they found that PC-1234 met their requirements, so we decided to patent the invention ourselves. The patent application was filed on August 12, 1992. U.S. Patent 5,457,165, titled "Encapsulant of Amine-Cured Epoxy Resin Blends" to Ralph D. Hermansen and Steven E. Lau on October 10, 1995.

12.14.2   DELCO ELECTRONICS CORPORATION—TWO PATENTS

Two patents were awarded covering essentially the same invention. The two patents have different titles, a somewhat different list of inventors, and different assignees. U.S. Patent 5,608,208 is assigned to both Delco Electronics Corporation and Hughes Aircraft Company, whereas U.S. Patent 5,185,498 is solely assigned to Delco Electronics Corporation.

12.14.3   U.S. PATENT 5,185,498

U.S. Patent 5,185,498 entitled **Circuit Assembly encapsulated with a PolyBd Urethane** was awarded to: Henry M. Sanftleben, Ralph D. Hermansen, Gary R. Shelton, Petrina L. Schnabel, Dennis T. Baird, and Douglas C. Smith on February 9, 1993. The patent application was filed on June 11, 1991. The invention described has significant cost and quality advantages over the materials and methodology used previously, namely, "Pour on sand." This POS process resulted in a filled encapsulant either by pouring the compound over sand already in the cavity or by adding sand to the encapsulant in the cavity. The POS method made it necessary to conformally coat the printed wiring board (PWB) to protect the circuitry during the potting process. The new method described in Patent 5,185,498 uses a special unfilled encapsulant, comprises polybutadiene polyol, polyoxypropylene triol, and a methylene bis (phenyl isocyanate).

This polyurethane was unique in standing up to the thermal cycling, salt spray testing to potted units without shrinking or cracking. Being unfilled, it was no longer necessary to conformally coat the PWB. Moreover, quality problems inherent with the POS method were eliminated.

12.14.4   U.S. 5,608,208

U.S. 5,608,208 entitled, "**Polybutadiene Urethane Potting Material**" was awarded to Henry M. Sanftleben of Delco Electronics and Ralph D. Hermansen of Hughes Aircraft on March 4, 1997. The assignee was Delco Electronics Corporation and Hughes Aircraft Company. The patent application was filed on February 2, 1994. Although patents 5,185,498 and 5,608,208 are very similar, the latter patent incorporates modifications learned in the 1991 optimization effort.

12.15   PRESENTATIONS

I was asked to be chairman of a session at the International SAMPE Electronics Conference held in New Jersey June 20–23, 1994. Henry M. Sanftleben (Hank) gave a paper on the air bag sensor potting development effort in my session.

KEYWORDS

- **Delco Electronics Corporation**
- **Flexipoxy potting**
- **potting compound**
- **Hacthane 110**

CHAPTER 13

REACTIVE HOT MELT CONFORMAL COATING MATERIALS

CONTENTS

At Delco Electronics, very high volumes of electronic units, such as radios, etc., are produced daily and in order to increase production, additional factory floor space and additional expensive processing equipment are required. In this chapter, the focus is on finding a way to conformally coat thousands of PWBs per day without adding new floor space or expensive curing equipment.

13.1   BACKGROUND—CONFORMAL COATINGS

Conformal coatings are commonly used to protect electronic assemblies from contamination, moisture, corrosive chemicals, and shock and vibration. The automotive industry and the defense/aerospace industry especially need conformal coatings to protect circuitry from hostile environments. Conformal coatings are most often composed of the following polymers: silicones, acrylics, polyurethanes, and epoxies. One problem with several commercially available conformal coatings is that they contain an organic solvent, which evaporates into the air during coating application. Environmental regulations outlaw many solvents and impose strict controls on the others. In addition to the solvent problem, conformal coatings require some kind of special processing be it metering and mixing of two component systems, dispensing, and curing the polymers in ovens, etc. It would be desirable to eliminate solvents and the need for special equipment and processing entirely.

13.2   THE HOT MELT CONFORMAL COATING CONCEPT

13.2.1   THE FAST PROCESSING OF HOT MELTS

Hank Sanftleben and Jim Rosson, Delco Electronics engineers, pondered this problem and came up with a brilliant solution. Their concept was to utilize hot melt conformal coatings in place of traditional materials. Hot melts need little in the way of application equipment; simply a small heating chamber to melt the hot melt and dispense it in its liquid state. The applied hot melt material quickly cools to a solid, allowing the next process to take place. This is the beauty of the concept, namely, more production can take place in the existing factory floor space. Moreover, equipment used to meter and mix two-component conformal coatings is no longer necessary because hot melts are already one component. Furthermore, conveyer ovens and the like are not needed because the hot melt quickly cools to a solid and the next process can begin. Furthermore, lengthy cure periods are eliminated to convert reacting liquid prepolymers to strong and useful cured polymers.

13.2.2 THE WEAKNESSES OF THERMOPLASTIC HOT MELTS

Traditional hot melts have one obvious problem. They are composed of lower molecular weight linear polymers in order to have the melting range be in a useful temperature range. In contrast, the majority of traditional conformal coating cures to become three-dimensional polymer networks. The difference between the two polymer types is that linear polymers will lose properties as they are heated near their melting temperature range and also may dissolve in certain solvents. Three-dimensional polymer networks do not melt and do not dissolve. Their useful service temperature range is therefore larger.

13.2.3 THE REACTIVE HOT MELT CONCEPT

The Delco engineers had a second important insight. Perhaps, a reactive hot melt material might be formulated such that it processes like a hot melt, but contains reactants, which gradually convert it into a three-dimensional polymer network. This conversion might not require special processing, but might occur during normal storage, during subsequent operations, or during actual use.

The hot melt conformal coating concept was also seen to be a solution to one of the problems of traditional conformal coatings. That problem is specific to integrated circuits (ICs). It is difficult or impractical to keep the conformal coating from filling the gap under the IC. However, the cured conformal coating under the IC is problematic because its thermal expansion exerts a stress on the soldered IC leads. The service life of ICs can be substantially reduced as a result. However, the hot melt conformal coating approach makes it possible to avoid under filling the IC component.

13.3   FEASIBILITY OF FORMULATING A REACTIVE HOT MELT MATERIAL

Hank and Jim wanted to know if our laboratory could accomplish such a formulation feat. In other words, could we develop a reactive hot melt material, which processed like a regular hot melt but continued to polymerize during storage times and/or during post processing conditions. It was outside our comfort zone, so we felt that we should begin with a feasibility study to get a feel for the difficulty of the task. Our formulating expertise has been with either epoxy or polyurethane compounds, so this was where we began.

13.3.1   POSTCURING TO A THERMOSET VIA MOISTURE-TRIGGERED CROSS-LINKING

One method of getting a gradual increase in cross-linking of these particular polymers is via moisture curing. One advantage of it is that moisture curing will not shorten the work-life of the hot melt. In the case of polyurethanes, providing for an excess of isocyanate groups in the formulation will result in gradual cross-linking as moisture gradually diffuses into the solidified hot melt material. The water molecules react with an isocyanate group to form a primary amine and carbon dioxide. The latter product diffuses out of the solidified hot melt material and escapes as a gas. The newly formed primary amine will instantly react with an isocyanate to form a urea linkage.

In the case of epoxy compounds, a similar effect can take place if there is an excess of epoxide groups and a ketimine has been added to the formulation. Ketimine curatives are basically blocked diamines. When water vapor diffuses into the solidified epoxy hot melt material, it reacts with the ketimine converting it into a diamine and a ketone. The ketone evaporates from the hot melt material, whereas the diamine reacts with available epoxide groups to cross-link the growing epoxy polymer.

Some of the factors affecting moisture curing are relative humidity, ambient temperature, molecular weight of the reactants, moisture permeability of the polymer, and others. Moisture cures will happen faster in humid areas than in deserts. Moisture cures will happen faster in summer than in winter. There is less likelihood of carbon dioxide bubbles forming in the coating if the molecular weight of the isocyanate or epoxide reactants is larger rather than smaller. Finally, moisture cures will be faster if the solidified hot melt material is hydrophilic rather than hydrophobic. Unfortunately, good dielectric properties usually come from hydrophobic polymers.

Reactive hot melt compounds must be packaged in hermetically sealed packaging until ready to use. Otherwise, they will cross-link to a non-meltable thermoset and be rendered useless.

13.3.2 POSTCURING TO A THERMOSET VIA HEAT-ACTIVATED LATENT CURATIVES

Besides the "moisture cure approach," there is the "latent curatives approach." Latent curative technology is probably better developed for epoxies than for polyurethanes. There are many commercially available latent curatives for epoxies, from which to choose. We already encountered this technology in Part II Chapter 9. Typically, the curing of the epoxy resins via latent curatives proceeds at a glacial pace at ambient temperatures. We know that some curing does occur because the viscosity of these compounds rises with increasing storage time. We also know that one-component epoxy compounds, based on latent curatives, can have long shelf lives on the order of six months or more.

These materials are cured by heating them to a temperature where the polymerization occurs in a reasonable time period, say 30 min to 2 h. Ideally, our hot melt material can be applied somewhere in the 40–90°C temperature range. We prefer that the latent curative remain reactively sluggish in this temperature range so that our hot melt application process is not negatively affected. On the other extreme, we would like to see reasonable cures in the 100–130°C range. Higher temperatures than this must be minimized for the sake of the electronic package.

13.4 TENTATIVE REQUIREMENTS FOR A RHMCC MATERIAL

The following material property requirements were defined to guide the formulating effort:

1. Processing: Hot melt application temperature should be between 125 and 200°C. Viscosity should be approximately 1000 centipoises for conformal coatings to flow properly, whereas adhesives could be up to 10,000 cps. Work-life of the molten reactive hot melts (RHM) should be about 3 h or more.
2. Material cost should be comparable with currently used materials.
3. The RHMCC should be flexible, not rigid or brittle.
4. The reactive hot melts should meet existing Delco Electronic requirements.

13.5   DEVELOPMENT OF A REACTIVE HOT MELT CONFORMAL COATING BASED UPON FLEXIBLE EPOXY TECHNOLOGY

13.5.1   BUILDING UPON FLEXIPOXY TECHNOLOGY

Conformal coatings should be flexible, even rubber-like to prevent damage to electronic components. We had already established a base of formulating knowledge of attaining rubber-like epoxy compounds. We even had created a small database on the compatibility of latent curatives with flexible epoxy resins. What we did not have was information on the compatibility of ketimines and flexible epoxy resins.

Secondly, a hot melt is a solid material at room temperature. Solid epoxy resins are commercially available, but the vast majority of them lead to a hard, brittle epoxy plastic. Could they be mixed with flexible epoxy resins and ketimines, a reactive hot melt conformal coating (RHMCC)? The following table lists the epoxy resins used in our feasibility effort.

TABLE 13.1   Description of Epoxy Resins Used in the Study

Tradename	Chemical type	Molecular weight	Epoxide equivalent	Repeating units
Epon 828	Bisphenol A	370–385	185–192	0.11–0.16
DEN 438	Novolac	650	180	3.6
DER 661	Bisphenol A	1000–1120	500–560	2.3–2.7
HPT 1071	See Note 1	640	150–170	None
MT 0163	See Note 2	770	185–200	None
NC 547	See Note 3	1800	600	No data

TABLE 13.1   *(Continued)*

Tradename	Chemical type	Molecular weight	Epoxide equivalent	Repeating units
NC514	See Note 4	350	190	None
DER 732	See Note 5	640	305–335	No data
Heloxy 505	See Note 6	1800	550–650	None
Heloxy 84	See Note 7	1950	620–680	No data

Note 1: N,N,N',N'-tetraglycidyl bis(aminiphenyl) p-diisopropyl benzene, a tetra-functional epoxy resin containing two tertiary amines, which may increase reactivity. The chemical structure follows:

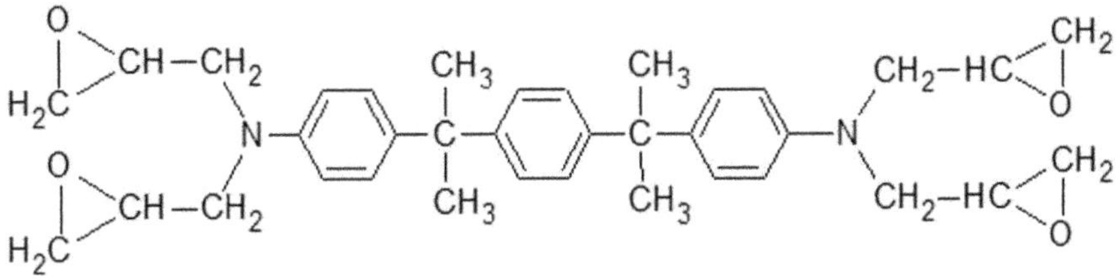

FIGURE 13.1   Chemical structure of Epon HTP 1071 epoxy resin.

Note 2: Huntsman took over this tetra-functional, phenol-based epoxy resin from Ciba. The Chemical structure follows:

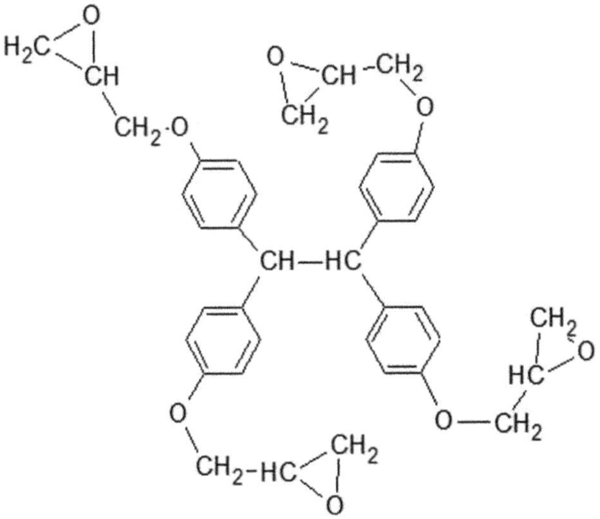

FIGURE 13.2   Huntsman MT 0163 epoxy resin.

Note 3: Cardolite tri-functional Novolac epoxy resin contains long, flexible, and pendant groups.

Note 4: Cardolite diepoxide, which has an eight carbon chain separating the aromatic groups.

Note 5: Dow diepoxide, which contains a poly(oxypropylene) chain separating the epoxides.

Note 6: Shell tri-epoxide based on castor oil.

Note 7: Shell tri-epoxide, with each poly(oxy-propylene) chain terminated with an epoxide group.

13.5.2   EPOXY RHMCC FEASIBILITY TESTING

Various epoxy resins were mixed with the indicated amount of Shell Chemical H-1 ketimine curative and allow to moisture cure for the indicated number of days. The epoxy resins fall into two categories: those intended for rigid polymer applications and those intended as flexibilizing co-resins. Solid epoxy resins were only found of the rigid type. All flexibilizing epoxy resins were liquids. Ketimine H-1 successfully converted both kinds of epoxy resins into cross-linked polymers.

Acetone resistance was used as a criterion for establishing that cross-linking had occurred in the compounds due to moisture curing. A rigid epoxy compound would be an insoluble solid, whereas a flexible compound would be a swelled rubber. Failure to moisture cure would show the hot melt dissolving in the acetone. Table 13.2 shows the results.

TABLE 13.2   Results of Epoxy Resin Moisture Cures

Epoxy resin name	Physical state	Epoxy resin type	phr H-1 ketimine*	Cure time, days	Acetone test results on cured polymer
Epon 828	Liquid	Rigid	27.4	6–8	Insoluble solid
DER 661	Solid	Rigid	9.9	>14	Soluble solid
DEN 438	Semi-solid	Rigid	29.2	7–10	Insoluble solid
HPT 1071	Solid	Rigid	32.5	7–14	Insoluble solid
MT 0163	Solid	Rigid	27.1	7–14	Insoluble solid
NC 547	Liquid	Flexible	8.7	2–4	Swelled rubber
NC514	Liquid	Flexible	14.9	4–7	Swelled rubber
DER 732	Liquid	Flexible	16.3	2–3	Swelled rubber
Heloxy 505	Liquid	Flexible	8.7	2–3	Swelled rubber
Heloxy 84	Liquid	Flexible	8.0	7–10	Swelled rubber
Heloxy 67	Liquid	Flexible	40.0	2–3	Swelled rubber

*Parts by weight added to 100 parts by weight epoxy resin

OBSERVATIONS

The rigid types became insoluble solids, including the initially liquid Epon 828. All of the flexible epoxy resins also cross-linked into rubbers. They did not liquefy as acetone was absorbed, they swelled. Elastomers, rubbers, and the like have the capacity to swell considerably as they absorb certain solvents. The polymer segments between cross-link nodes are typically only a fraction of their stretched out length due to continuous wriggling. As more and more solvent is absorbed, the polymer segments stretch farther and farther until the swelled rubber finally splits open. Rigid cross-linked polymers, on the other hand, have little to no capacity to swell.

13.5.3 SPECIAL TEST PROCEDURE FOR EVALUATING MOISTURE CURE RATES

A mold with a cylindrical cavity was made such that the diameter is 1 cm and the length is 4 cm. The reactive hot melt is heated to a temperature where it becomes a pourable fluid and is poured into the mold. Upon cooling, the solidified sample is removed from the mold and is allowed to absorb moisture from ambient air. Moisture from the air permeates the specimen from the outside inward. Consequently, the full cure of the sample also progresses from the outer surface inward. Periodically, a section is cut off and soaked in methyl ethyl ketone (MEK). The uncured portion of the specimen will dissolve in the MEK. Preliminary trials show that a ring of cured material remains after the MEK exposure. The thickness of the ring increases as the period of moisture curing increases. The thickness of the cured ring is measured using calipers. The rate of cure can be determined from a plot of ring thickness versus time.

13.5.4 REACTIVE EPOXY HOT MELTS CONTAINING FILLERS

Delco Electronics was interested in other hot melt applications where fillers would be included in the formulations. Several advantages might be realized using fillers in the reactive epoxy hot melt formulation: (1) incompatible ingredients are held together, (2) the heat capacity of the filler would extend the melt phase, (3) fillers reduce tackiness, and (4) raw material cost is reduced. Table 13.3 shows formulations for three filled epoxy reactive hot melts. Table 13.4 shows their properties.

TABLE 13.3   Formulation for Filled Reactive Hot Melts

Ingredient	Function	FERHM-1	FERHM-2	FERHM-3
Epon 1004F	Solid epoxy resin	70 pbw	55 pbw	22 pbw
Heloxy 84	Liquid flexible epoxy resin	25 pbw	45 pbw	None
DER 732	Liquid flexible epoxy resin	None	none	12 pbw
LHT-34	Liquid long chain triol	5 pbw	5 pbw	None
Ketimine H-3	Moisture-activated diamine	12 pbw	12.6 pbw	6.3 pbw
Alumina powder	Filler	300 pbw	300 pbw	63.5 pbw
Aer-O-Sil R972	Thixotrope	None	None	2.5 pbw

Note: The chemical structure of Epon 1004F is as follows:

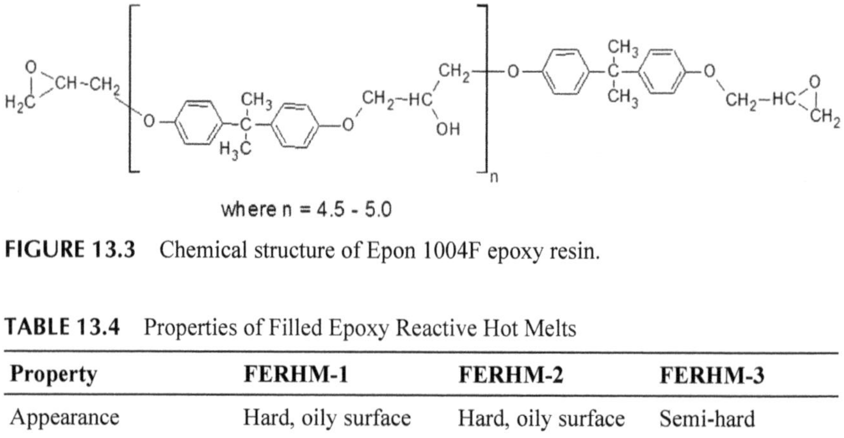

where n = 4.5 - 5.0

FIGURE 13.3   Chemical structure of Epon 1004F epoxy resin.

TABLE 13.4   Properties of Filled Epoxy Reactive Hot Melts

Property	FERHM-1	FERHM-2	FERHM-3
Appearance	Hard, oily surface	Hard, oily surface	Semi-hard
Melt temperature, °C	100	90	90
Moisture cure rate *	30 mils in 4 days	60 mils in 4 days	60 mils in 4 days
Melts at 250°C **	No	No	No

* Thickness which does not dissolve in acetone due to cross-linking.

** Cured reactive hot melt no longer melts.

13.5.5 *REACTIVE EPOXY HOT MELT COMPOUNDS BASED ON JEFFAMINE ED-4000 DIAMINE*

13.5.5.1 *USING SOLID DIAMINES TO FORMULATE A HOT MELT*

Instead of using solid epoxy resins in order to end up with a solid reactive hot melt, one could perhaps use solid diamines. If they also contributed to the flexibility of the RHMCC due to having a flexible polymer chain between terminal amine groups so much the better. This approach also allows us to use liquid, flexibilized, diepoxide resins to build the epoxy-terminated solid prepolymer. The Jeffamine polyether amines from Huntsman Company have several candidates, which meet the need. ED-4000, being a solid, was selected for evaluation with different liquid epoxy resins.

The general structure for the Jeffamine ED series is shown below.

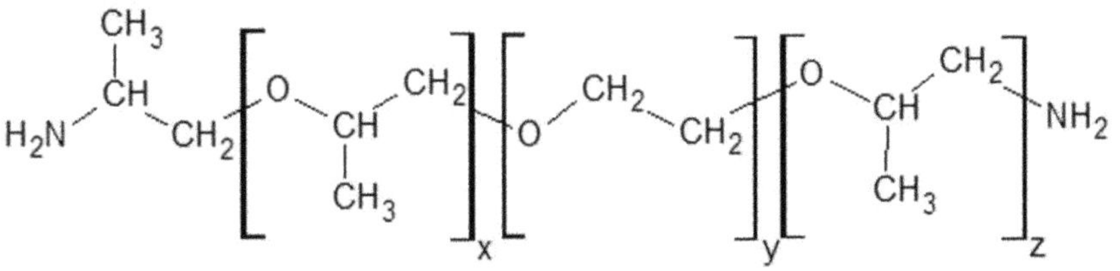

FIGURE 13.4 Structure for the Jeffamine ED series.

13.5.5.2 *MAKING AN EPOXIDE-TERMINATED PREPOLYMER*

Primary diamines, such as Jeffamine ED-4000, have a total of four active hydrogens. Four different epoxide groups can chemically attach to one diamine molecule. The flexible epoxy resins, which we want to attach to the diamine, have two or more epoxide groups per molecule. We need to combine four molecules of epoxy resin to each one molecule of diamine. The resultant prepolymer structure will have at least four free epoxide groups on it.

The procedure for preparing the ED-4000 prepolymer reactive hot melt compound is as follows:

(1) Melt the ED-4000 at an elevated temperature (e.g., 80°C),

(2) Add liquid polyepoxide resin at a ratio of 4 moles of polyepoxide per 1 mole of ED-4000,

(3) Combine the reactants and continuously stir them at 150°C in a closed metal container for 1 h.

(4) Reduce the temperature and add the curative with stirring. Either a latent curative or a ketimine for moisture curing could be added.

(5) Mix under vacuum for 30 min,

(6) Transfer the reactive hot melt to hermetically sealed metal cartridges or in heat-sealed foil-lined pouches.

13.5.5.3   TEST RESULTS FOR ED-4000 REACTIVE HOT MELTS

Table 13.5 shows the solid meltable prepolymers made with ED-4000 and liquid flexible epoxy resins. Table 13.6 shows that reactive hot melts using latent curatives with these prepolymers yielded rubber-like compounds with the desired melting points. Table 13.7 shows that reactive hot melts using different ketimines also yielded rubber-like compounds with the desired melting points.

TABLE 13.5   Composition of Epoxy-Terminated Prepolymer

Prepolymer ID	Epoxy resin	Diamine	Description of prepolymer
EPP-1	Heloxy 67	ED-4000	Waxy solid
EPP-2	Heloxy 84	ED-4000	Soft waxy solid
EPP-3	DER 736	ED-4000	Waxy solid
EPP-4	Cardolite NC-514	ED-4000	Waxy solid
EPP-5	Epon 828	ED-4000	Waxy solid

TABLE 13.6   Reactive Hot Melt Compounds Containing Latent Curatives

ID	Prepolymer ID, pbw	Aer-O-Sil R972 pbw	AH-122 pbw	Melting point, °C	Durometer, Shore A
RHM-L1	EPP-1, 24.0	0.5	6.2	39	82
RHM-L2	EPP-2, 25.0	0.5	3.5	45	45
RHM-L3	EPP-3, 25.4	0.5	5.9	41	75
RHM-L4	EPP-4, 24.0	0.5	4.1	40	36
RHM-L5	EPP-5, 25.4	0.5	5.9	40	40

Note: Ajicure AH-122 is an aliphatic dihydrazide, which serves as a latent curative.

TABLE 13.7   Reactive Hot Melt Compounds containing Moisture-Activated Curatives

ID	Prepolymer ID,	Ketimine name, pbw	Composition, prepolymer, pbw, Ketimine, pbw	Melting point, °C	Durometer, Shore A
RHM-K1	EPP-1	H-1	25.0–2.6	39	82
RHM-K2	EPP-1	H-2	25.0–2.8	39	82
RHM-K3	EPP-1	H-3	25.0–5.0	39	82
RHM-K4	EPP-3	H-1	25.0–2.4	41	78
RHM-K5	EPP-5	H-1	25.0–2.4	40	85
RHM-K6	EPP-4	H-1	25.0–1.8	40	80

13.5.5.4   CONCLUSIONS ABOUT ED-4000-BASED EPOXY RHMS

The Jeffamine ED-4000 reactive hot melts, once cured, only yielded an insulation resistance of $1–10 \times 10^8$ ohms. It was surmised that the high frequency of ether groups in the ED-4000, makes it very hydrophilic, which relates to the very poor electrical resistivity. Conformal coatings should have values three or more decades higher, so these fail that test.

Too bad that there are not any Jeffamine diamines based on PTMEG or poly(oxy-tetra-methylene) structure. I predict that they would give better electrical properties than ED-4000. Even longer polymethylene chain diols would be better. Be aware that it is the stereoregularity of an ethylene oxide polymer, which gives it its solid nature.

Despite the poor volume resistivities of ED-4000 RHMCCs, the experiment was very useful. We used the only available solid diamine to prove the solid diamine concept. We formulated a flexible epoxy RHMCC using liquid flexible epoxy resins and a solid diamine curative. Both the moisture-curing mechanism and the latent cure mechanism have been demonstrated to work.

13.6   DEVELOPMENT OF A REACTIVE HOT MELT CONFORMAL COATING BASED UPON POLYURETHANE TECHNOLOGY

13.6.1   BACKGROUND

One-component, moisture-curing polyurethane coatings have been commercially available for decades and are principally used as coatings. They typically are isocyanate-terminated prepolymers, prepared by reacting a diisocyanate with a diol or triol. The molecular weight of the prepolymer can be predetermined by adjusting the ratio of NCO to OH groups. The formulated coating may contain organic solvents to reduce viscosity and other additives to produce a smooth level coating. For outdoor applications, UV absorbers and antioxidants may be added to the coating formulation to extend its life. These coatings are typically applied only a few mils thick. Moisture, absorbed into the coating from the air, causes the prepolymer to link with itself via urea linkages. The water molecule converts an isocyanate group into a primary amine, which, in turn, reacts with a second isocyanate group to form the urea linkage. Technically speaking, the moisture-cured prepolymer is now both a polyurethane and a polyurea polymer.

13.6.2 FORMULATING CONSIDERATIONS

These commercial coatings are quite different from a typical hot melt. The existing moisture-curing coatings are low viscosity liquids, whereas hot melts are firm, solid materials. In order to formulate a moisture-curing, polyurethane hot melt material, we have to find ingredients, which produce a meltable solid and have terminal isocyanate groups in the molecule. Ideally, we would have little to no cross-linking reactions occur during the process of making an isocyanate-terminated prepolymer. Excessive cross-linking might render the prepolymer unmeltable or unflowable at elevated temperatures above its melt temperature. The functionality of the reactants is important here. If we limit the functionality of the isocyanates and polyols to only bifunctional, we should be able to build linear polymeric molecules. Thus, we wish to select diisocyanates and diols as starting materials to accomplish this goal.

Next, another variable at our control is the ratio of isocyanate (NCO) groups to hydroxyl (OH) groups. If we select a ratio of two isocyanates for each hydroxyl group, we should obtain the smallest isocyanate-terminated prepolymer, whereas if we select a ratio of one isocyanate to each hydroxyl group, we are likely to obtain a linear polymer of high molecular weight and having no free isocyanate groups available for subsequent moisture curing. So the range between a NCO/OH ratio of 2:1 and 1:1 is our area of interest. Intermediate NCO/OH ratios of say 1.75:1 would produce isocyanate-terminated prepolymers with sufficiently high molecular weight to be solid at room temperature.

13.6.3   A SOLID ISOCYANATE-TERMINATED PREPOLYMER BECOMES A RHMCC

As I reviewed the formulation development work that we did toward a reactive, polyurethane, hot melt conformal coating, I was impressed at what we had accomplished. Moreover, the science of some experiments was something I really wanted to share. The only solid reactive polyurethane mentioned in this patent is what I will refer to as URHM-1. URHM-1 was the first prepolymer made in our lab, which was a solid material. It was the combination of 1.8 moles of Henkel DDI-1410 to 1.0 mole of 1,4-butanediol. The prepolymer melted between 50 and 75°C. Lap shear specimens were evaluated with the following results:

Initial: 20 psi,
After 10 days moisture cure: 160 psi.

13.6.4   REACTIVE HOT MELT CONFORMAL COATINGS BASED UPON URHM-1 MODIFICATIONS

The longer polymethylene sequence in the diol chain, the more hydrophobic the polymer will be and perhaps the more dielectric it will be. These diols are readily available so we decided to look into prepolymers based on longer diols. The diol descriptions are shown in Table 13.8.

TABLE 13.8   Description of Polyurethane Hot Melt Prepolymers

Name of prepolymer	Diol	Melting point, °C	Solid formed?
URHM-1	1.4-Butane diol		Yes
URHM-2	Six carbon diol	42	Yes
URHM-3	Eight carbon diol	60	Yes
URHM-4	10 carbon diol	72	Yes
URHM-5	12 carbon diol	83	Yes
URHM-6	Cycloaliphatc diol	32	No*

*URHM-6 did not become a solid so it was dropped from the test program.

Table 13.9 shows the processing characteristics of these prepolymers.

TABLE 13.9　Processing Characteristics of Polyurethane Reactive Hot Melts

Property	URHM-1	URHM-2	URHM-3	URHM-4	URHM-5
Melting point	53°C	60°C	64°C	73°C	77°C
Viscosity (cps)					
at 50°C	Solid	Solid	Solid	Solid	Solid
At 75°C	1350	1500	1750	Semisolid	Solid
At 100°C	300	450	400	900	575
At 125°C	165	200	175	350	250
Sprayability temperature	95°C	105°C	100°C	125°C	110°C

All of the candidates melt in a desirable low temperature range and can be spray applied as a conformal coating between 95 and 125°C. Table 13.10 shows the physical properties of the RHMCCs.

TABLE 13.10　Physical Properties of Polyurethane Reactive Hot Melts

Property	URHM-1	URHM-2	URHM-3	URHM-4	URHM-5
Durometer	48 D	52 D	55 D	57 D	58 D
T_g, °C	−8.5	−14	−14	−13	−13
CTE below T_g	110	112	109	113	115
CTE above T_g	163	180	196	187	177
Softening point	43°C	43°C	48°C	49°C	55°C
Melting point	>160°C	>160°C	>160°C	>160°C	>160°C
Adhesion	Excellent	Excellent	Excellent	Excellent	Excellent

Comments:

The cured reactive hot melts all have higher hardness than desired, but may have enough flexibility to function as conformal coatings. As the chain length of the diols increased, the hardness also increased. Perhaps, a flexibilizing group, such as an ether linkage in the center of the diol chain might soften the compound sufficiently. The glass transition temperature (T_g) decreased between four and six carbon diols and then stayed constant as the number of carbon atoms increased in the diol.

Interpreting the coefficient of thermal expansion (CTE) above and below the glass transition temperature versus diol chain length is not a precise task. The CTE values are not hard numbers but up to how someone fits a straight line through a series of points on a graph. I would say that all of the CTE values are probably within experimental error and there is no discernible trend. The softening point of the cured reactive hot melts appears to increase with increasing chain length of the diol, whereas the uncured reactive hot melts melted between 53 and 77°C, the cured versions did not melt at the test temperature of 160°C. The adhesion of the reactive hot melt was excellent. This is an important property for a conformal coating.

13.6.4.1   A TEST FOR DETERMINING INSULATION RESISTANCE

Good dielectric properties are so important to a RHMCC that we developed a test procedure to evaluate insulation resistance under different conditions. Insulation resistance is an empirical test, but correlate well with volume resistivity. Here are the details of the test:

Y-pattern specimens of a type specified in MIL-I-46058 were used in the evaluation of the reactive hot melt conformal coatings. The 1.5 × 3 inch G10 epoxy glass laminates had copper traces in the Y pattern. The specimens were spray coated and allowed to moisture cure for 2 weeks. A moisture resistance test was performed by subjecting the specimens to cycling from 25 to 65°C at 95% relative humidity. A dwell time of 3 h at 65°C was held before dropping again to 25°C.

Insulation resistance readings were taken at the following times:

1.   Immediately after the coating was applied and allowed to cool.
2.   After 2 weeks cure of the coating at 25°C and 60% RH.
3.   Immediately prior to moisture resistance testing.
4.   During moisture resistance testing on cycles 1, 4, 7, and 10. Readings are taken in the middle of the 65°C dwell period.
5.   Upon recovery from the moisture resistance testing.
6.   After completion of the accelerated hydrolytic stability test. This test consists of 48 h at 250°F and 15 psig steam pressure in an autoclave.

13.6.4.2   THE INSULATION RESISTANCE OF OUR POLYURETHANE RHMCC CANDIDATES

In Table 13.11, we compare the insulation resistance of the unreacted hot melt with the moisture-cured hot melt.

Table 13.11   Insulation Resistance of Polyurethane Reactive Hot Melts

Condition	URHM-1	URHM-2	URHM-3	URHM-4	URHM-5
No cure	2.2×10^{12} ohm	4.8×10^{12} ohm	6.2×10^{12} ohm	1.0×10^{13} ohm	1.2×10^{13} ohm
14-day cure	2.0×10^{12} ohm	5.0×10^{12} ohm	6.0×10^{12} ohm	1.0×10^{13} ohm	1.6×10^{13} ohm

Comments:

There seems to be no significant improvement in the reactive hot melt coatings due to moisture curing. However, there does seem to be strong correlation between insulation resistance and number of carbon atoms in the diols. Our theory of a correlation between hydrophobicity and dielectric properties seems to be supported by the evidence.

In Table 13.12, the candidates are measure through the stages of the moisture resistance test.

TABLE 13.12   Insulation Resistance of Polyurethane Reactive Hot Melts in Moisture Resistance Test

Condition	URHM-1	URHM-2	URHM-3	URHM-4	URHM-5
Initial	2.2×10^{12} ohm	5.0×10^{12} ohm	6.2×10^{12} ohm	1.1×10^{13} ohm	1.4×10^{13} ohm
Cycle 1	1.1×10^{9} ohm	3.0×10^{9} ohm	3.2×10^{9} ohm	3.5×10^{9} ohm	7.5×10^{9} ohm
Cycle 10	8.0×10^{8} ohm	2.0×10^{9} ohm	2.2×10^{9} ohm	2.5×10^{9} ohm	5.0×10^{9} ohm
24 h later	3.5×10^{11} ohm	1.5×10^{12} ohm	1.4×10^{12} ohm	2.0×10^{12} ohm	2.2×10^{12} ohm

Note: Although I had the data for cycle numbers 1, 4, 7, and 10, I only reported cycles 1 and 10 in the table above for simplicity. The samples lost three decades of insulation resistance during the first cycle and only slowly declined thereafter.

Comments:

Again, there does seem to be strong correlation between insulation resistance and number of carbons in the diols. However, the most interesting thing in the data is how rapidly the samples returned to their higher insulation resistances after recovering at ambient conditions.

In Table 13.13, the candidates are subjected to the hydrolytic stability test.

TABLE 13.13 Insulation Resistance of Polyurethane Reactive Hot Melts in Hydrolytic Stability Test

Condition	URHM-1	URHM-2	URHM-3	URHM-4	URHM-5
Before	3.5×10^{11} ohm	1.5×10^{12} ohm	1.4×10^{12} ohm	2.0×10^{12} ohm	2.2×10^{12} ohm
After	6.0×10^{9} ohm	7.5×10^{10} ohm	2.0×10^{11} ohm	2.9×10^{11} ohm	3.6×10^{11} ohm

Note: The samples from the moisture resistance test were immediately put into the hydrolytic stability test after their 24-h recovery. The hydrolytic stability test consists of 48-h at 250°F and 15 psig steam pressure.

Comments:

Sample URHM-1 fared the worst in this severe test. It softened enough to deform during the test but was not permanently damaged. It was a firm solid again upon recovery. Again, there does seem to be strong correlation between insulation resistance and number of carbon atoms in the diols. URHM-1 is nearly two decades lower than URHM-5 after the 48-h steam exposure.

13.6.5 EVALUATION OF DIOL X AS A PATH TO LOWER HARDNESS REACTIVE HOT MELTS

The results of the series URHM-1 through URHM-5 were very encouraging. However, a softer, lower durometer reactive hot melt compound would assure that sensitive component would not be damaged during thermal cycles.

13.6.5.1   INGREDIENT VARIATION STUDY DESCRIPTION

Diol X is a liquid hydroxyl-terminated polymer with a rubber-like polymer chain. It has a molecular weight >1000 and a low glass transition temperature <50°C. The following experiment is designed to replace the 1,4-butane diol in URHM-1 formulation in steps and appraise the benefits.

Table 13.14 shows the formulation of the trial prepolymers.

TABLE 13.14   Composition of Diol X Replacement Reactive Hot Melt Compounds

Name	Percent of butane diol in diol blend	Percent of Diol X in diol blend	Solid product?
URHM-1	100	0	Yes
URHM-6	90	10	Yes
URHM-7	80	20	Yes
URHM-8	70	30	Yes
URHM-9	0	100	No

Note: URHM-9 was dropped from further evaluation.

13.6.5.2   PROCESSING CHARACTERISTICS OF INGREDIENT VARIATION STUDY CANDIDATES

Table 13.15 shows the processing characteristics of the candidates.

TABLE 13.15   Processing Properties of Diol X Replacement Reactive Hot Melt Compounds

Property	URHM-1 (6 months old)	URHM-1 (freshly made)	URHM-6	URHM-7	URHM-8
Melting point	53°C	60°C	64°C	73°C	77°C
Viscosity, cps					
at 50°C	Solid	Solid	Solid	Solid	Solid
at 75°C	1400	1175	1275	1375	1250
at 100°C	380	350	400	325	400
at 125°C	165	150	160	150	165
Sprayability	95°C	95°C	100°C	95°C	100°C

Comments:

The hot melt melting points increased with increasing diol X in the diol blend. Viscosity versus temperature behavior and the sprayability temperature was virtually the same for all candidates.

13.6.5.3 THE CURED CHARACTERISTICS OF INGREDIENT VARIATION STUDY CANDIDATES

Table 13.16 shows the properties of these RHM candidates after curing to thermosets.

TABLE 13.16   Properties of Diol X Replacement, Cured RHM Compounds

Property	URHM-1	URHM-6	URHM-7	URHM-8
Durometer	46 D	38 D	40 D	42 D
T_g, °C	−8.5	−25	−20	−23
Softening point, °C	43	30	30	40
Melt below 160°C?	No	No	No	No
Adhesion	Excellent	Excellent	Excellent	Excellent

Comments:

(1) There seems to be a trend of increasing hardness with increasing Diol X content.
(2) The glass transition temperature, Tg, of the reactive hot melts was significantly lowered due to the addition of Diol X to the diol blend. Ideally, the Tg should have decreased with increasing Diol X content. Instead, no clear trend is indicated.
(3) After moisture curing, all of the hot melts no longer melted at the 160°C test temperature.
(4) Adhesion of the cured coatings was excellent.

13.6.5.4 THE ELECTRICAL PROPERTIES OF INGREDIENT VARIATION STUDY CANDIDATES

The insulation resistance of URHM-1 and URHM-6, 7, and 8 fell in the range 2.0–2.6×10^{12} ohm prior to moisture curing, but barely changed after a 2-week cure. In the moisture resistance test, the same candidates lost three decades of insulation resistance during the first cycle, and stayed in the 10th range through cycle 10. They all recovered by two decades within 24 h at ambient conditions. The accelerated hydrolytic stability test was immediately run on these specimens and showed no degradation.

The dielectric constant of the same candidates fell in the 1.6–2.3 range, the dissipation factor fell in the 0.079–0.112 range at 1 kHz, the 0.045–0.061 range for 10 kHz, and the 0.023–0.031 range at 100 kHz. The average dielectric strength for each candidate fell in the range 588–796 V/mil.

13.7 CONCLUSIONS

Reactive hot melt compounds for use as conformal coatings can be developed from either flexible epoxy technology or polyurethane technology. The compounds, which we developed, could be applied using spray equipment for hot melts at a low enough temperature to be safe for electronic components. This offers the advantage of faster processing as no ovens for curing are needed. The melted hot melt quickly solidifies and the part can move to the next operation.

Although flexible epoxy RHMCCs were proven feasible using latent curatives, we prefer the moisture cure approach. Ketimines added to the epoxy hot melts make the moisture cure approach possible. Flexible epoxy hot melts were proven feasible by two methods: (1) blending solid rigid epoxy resins with liquid flexible epoxy resins, or (2) mixing liquid flexible epoxy resins with solid diamine curatives.

Polyurethane RHMCCs can be made by reacting a diisocyanate about 2 moles to one of a diol or blend of diols where the resultant isocyanate-terminated prepolymer is a meltable solid. Moisture curing to cross-link the reactive hot melt proceeds via a reaction between ingressing water vapor and free isocyanates.

13.8 REACTIVE HOT MELT PATENTS

13.8.1 U.S. PATENT 5,510,138

U.S. Patent 5,510,138 entitled "Hot Melt Conformal Coating Materials" was awarded to Henry M. Sanftleben and James M. Rosson of Delco

Electronics Corporation and Ralph D. Hermansen of Hughes Aircraft Company on April 23, 1996. The assignees were Delco Electronics Corporation and Hughes Aircraft Company. The patent application was filed on May 24, 1994. The abstract of the patent describes a conformal coating material for the surface of electronic assemblies, where the conformal coating is a hot melt. It is solid or semisolid at room temperature and it will become a flowable liquid at an elevated temperature for application. Upon cooling, it becomes a solid again. The invention covers both hot melt conformal coatings and reactive hot melt conformal coatings. In the latter case, the reactive hot melt converts into a cross-linked, unmeltable polymer after application and at temperatures not exceeding the application temperature.

13.8.2   U.S. PATENT 5,965,673

U.S. Patent 5,965,673 entitled "Epoxy-Terminated Prepolymer of Polyepoxide and Diamine with Curing Agent" was awarded to Ralph D. Hermansen and Steven E. Lau on October 12, 1999 and on April 23, 1996, respectively, and assigned solely to Raytheon. The patent application had been filed on April 10, 1997. I had fully retired before this 1999 patent award, whereas U.S. 5,510,138 is focused on hot melt conformal coatings and includes both reactive and non-reactive hot melts, both epoxy and polyurethane reactive hot melts in its scope; U.S. 5,965,673 focuses on the chemistry of formulating reactive solid epoxies alone. Moreover, the applicability of the solid epoxy prepolymers is highly generalized. It could apply to conformal coatings, adhesives, or a myriad of other uses.

KEYWORDS

- **conformal coatings**
- **hot melts**
- **integrated circuits**
- **RHMCC**
- **URHM**

CHAPTER 14

SOLDER JOINT LEAD ENCAPSULATION

CONTENTS

14.1   WHAT IS A SOLDER JOINT LEAD ENCAPSULANT?

Electronic circuit assemblies often must survive under hostile operating conditions. Such conditions often exist in automotive or military and aerospace applications. Such assemblies often employ surface mount integrated circuit packages, which attach to the printed wiring substrate by soldering their leads. It is the failure of those soldered leads that concern us. Outdoor applications typical for autos or military equipment may see many hundreds of thermal cycles in which differential expansion forces can apply stress to the soldered leads. Such stressing can cause fatigue and fracturing of the soldered leads. Applications involving vibration may hasten the solder fatigue.

The problem is often exacerbated due to the conformal coatings, commonly used. Conformal coatings are usually flexible materials, which have a high coefficient of thermal expansion. When it is under the leads, its greater expansion rate than the solder and leads creates a tensile stress on the soldered joint. Low profile ICs are especially vulnerable because their leads cannot deform as much as higher profile leads.

One solution to the problem is use a solder joint lead encapsulant (SJLE) to bond the IC firmly to the printed wiring board. These encapsulants are typically epoxy compounds, containing a filler to match the CTE to that of solder, and formulated to have a glass transition temperature higher than the upper service temperature.

14.2   THE SPECIFIC PROBLEM

Hank Sanftleben and Jim Rossen of Delco Electronics requested that we assist them in developing a better solder joint lead encapsulant (SJLE). Teresa Lindley of Delco Electronics was in charge of the SJLE improvement project.

The SJLE, which they were currently using, has several serious problems:

(1) It requires a cure schedule of 3 h at 150°C or 6 h at 125°C. These times are vastly too long for the high production rates at Delco Electronics.
(2) Suitable conveyor ovens cost $100,000 or more.
(3) A prolonged cure temperature of 150°C is harmful to components of the electronic package, such as electrolytic capacitors. Moreover,

the heat exposure reduces the solderability of conductors due to formation of intermetallics.

(4) It cures upon application to a non-repairable state.

14.3 THE REACTIVE HOT MELT SJLE CONCEPT

Hank Sanftleben and Jim Rossen had a bold idea for solving the problems just stated. They visualized an epoxy hot melt SJLE, which can be applied as a thin bead across the IC leads using hot melt processing equipment. The application temperature is ideally 90–100°C. The hot melt SJLE material solidifies upon cooling to room temperature and the assembly can proceed to the next operation, eventually wave soldering. The hot melt epoxy is unreacted on application and removal of the IC at this stage is simply done by remelting the hot melt SJLE material and pulling the component off.

The new SJLE material is not simply a hot melt, but is a reactive hot melt. It is through the polymerization reaction that the SJLE material develops a cross-linked polymer structure and a high glass transition temperature. Once converted, the material is no longer meltable but is a strong rigid material to a temperature higher than the upper service temperature of the electronic unit. The heat exposure needed to trigger and complete the conversion would ideally be provided in post application steps. For example, the heat exposure of the wave-soldering step might be adequate to cure the SJLE. The investment in new curing ovens would be avoided.

14.4 OUR EARLIER WORK ON THE SJLE PROBLEM

Before there was an emphasis on the reactive hot melt approach, our lab at Hughes attempted to solve the cure temperature part of the problem using conventional ingredients. Our goal was to develop a solder joint lead encapsulant (SJLE), which cures and develops its full properties at significantly lower temperatures than 150°C. A cure temperature of 120°C was considered ideal. Reduction of cure time was also a priority.

We did a quite extensive analysis of epoxy resin/curative combinations to find the best ingredients. Then using an iterative process of "formulate, test, and reformulate," we developed a solder joint lead encapsulant designated SJLE 6141-A. This particular development effort did not get patented, so the formulation information is not in the public domain. However, the success of the effort is reflected in the properties of SJLE 6141-A shown in the following table:

TABLE 14.1 Cure Properties of SJLE-A Solder Joint Lead Encapsulant

Cure temperature	Cure time	Tg	CTE below Tg	CTE above Tg
120°C (248°F)	5 min	77°C	25.2 ppm/°C	77.7 ppm/°C
130°C (266°F)	5 min	135°C	25.2 ppm/°C	78.4 ppm/°C
130°C (266°F)	30 min	155°C	24.5 ppm/°C	76.4 ppm/°C

Delco Electronics engineers evaluated SJLE 6141-A on actual hardware.

14.5 PROPERTY OBJECTIVES FOR OUR REACTIVE HOT MELT SJLE MATERIAL

Table 14.2 below summarizes the property requirements for our reactive hot melt SJLE.

TABLE 14.2 Formulating Objectives for the RHM SJLE Material

Property	Reason	Requirement
Cools to a solid again at room temperature	To be able to proceed to next operation ASAP	Resolidify in minutes
Work-life as a hot melt	Typical time needed for hot melt delivery equipment	4 h or more
Application temperature		45°C–90°C
Viscosity at application temperature	In order to flow into crevices	5000 cp. max. 1000 cp. ideal
Cure schedule To cross-link the hot melt	Low heat exposure to prevent damage to IC leads and solder	Under 150°C, 120°C ideal. Quicker cures times if cure temperature is higher >120°C
Glass transition temperature	Must remain rigid at highest service temperature (i.e., 125°C)	>135°C
CTE of SJLE <T_g	Match solder CTE	25–35 ppm/°C
Repairability	Replace IC if necessary. Note: solder melts at 180°C	Be a hot melt material

14.6 TECHNICAL APPROACH

Most of the formulation development so far in this book has had flexible or rubber-like end products as a goal. For this particular application, we need to think more in terms of traditional epoxy formulation, where the end product is hard and rigid and the glass transition temperature is as high as possible. We also want to borrow from what we learned in the previous chapter concerning the development of an epoxy reactive hot melt conformal coating. We wanted to utilize the fast processing of hot melt technology, yet attain the thermoset characteristics of reactive hot melts.

Numerous solid epoxy resins are commercially available. This made our task of formulating a hot melt easier. Liquid epoxy resins can be co-blended to lower the melting point of the hot melt. Cross-linking to a thermoset can be accomplished via heat-activated latent curatives or via moisture-activated ketimines. The preferred curative approach for the SJLE is via latent curatives. Latent curatives are also solids and facilitate the making of a hot melt solid. The fully cross-linked SJLE would no longer melt or be dissolved by solvents. The cross-linking process also raises the glass transition temperature to the necessary value where the SJLE has adequate stiffness and strength at the highest service temperature to reinforce the leads.

In order to match the coefficient of thermal expansion to that of solder, glass, or mineral filler is added to the polymer at approximately 40 weight percent. Upon curing, the resultant composite has a lowered coefficient of thermal expansion. Glass microspheres can serve this function with minimal raising of viscosity at melt temperature. It is known that moisture ingress can lower epoxy adhesion to a glass surface. Thus, silane coupling agents were used to help maintain good adhesion of the epoxy matrix to the glass microspheres.

If the addition of glass microspheres raises the viscosity at application temperature excessively, there is a method to counteract this problem called bi- or tri-modal packing. To illustrate this concept on a macro scale, imagine a stack of cannon balls. Perhaps a tennis ball or ping pong ball would fit in the large gaps between cannon balls. Perhaps, BBs would fit in the spaces between the cannon ball and tennis ball. The same concept works on a micro scale. Glass microspheres can be obtained in a range of sizes; 40–100 micron glass spheres blend nicely with 8–30 micron glass spheres. Finally, glass spheres less than 5 microns will fit in the remaining spaces.

14.7 EXPERIMENTAL

14.7.1 FORMULATIONS OF SUCCESSFUL CANDIDATES

Borrowing from the research done on the reactive hot melt conformal coating project and after a trial and error period, four candidates were found to meet the requirements. Table 14.3 shows their formulations.

TABLE 14.3 Reactive Hot Melt Solder Joint Lead Encapsulants

Ingredient	Formulation A	Formulation B	Formulation C	Formulation D
Huntsman Tactix 742			19.7	19.7
Shell Epon 836	20.3	22.3		
Shell Epon 825	1.9	0.9	2.55	
Curezol C17Z		0.9		1.0
Ajinomoto Ajicure PN-23	1.9		2.55	
A glass spheres (29 micron)	75.8	75.8	75.1	75.9
Coupling agent A-187	0.1	0.1	0.1	0.1

Comments:

The first two epoxy resins (Tactix 742 and Epon 836) are solids at room temperature, but liquify at moderate temperature. Tactix 742 is a trifunctional epoxy resin, usually used where high temperature stability is needed. Epon 836 is a DGEBA-type epoxy resin. The addition of a small amount of Epon 825 liquid epoxy resin lowers the hot melt application temperature.

The presence of latent curing agents (Curezol C17Z and Ajicure PN-23) in the formulations allows the hot melt to gradually transform into a cross-linked, unmeltable polymer with each excursion at elevated temperature. Curezol C172 is an imidazole, is solid at room temperature, melts between 86 and 91°C, and has a long latency period in the reactive hot melt compound. Ajicure PN-23 is another latent curative. Chemically, it is an amine adduct with epoxy resin.

The glass microspheres constitute the major ingredient by weight. Considering that the glass is more than twice as dense as the polymers, the relationship by volume is much tighter. The coupling agent helps the epoxy polymer to adhere more permanently to the glass microspheres. The glass/epoxy composite has an expansion coefficient closer to that of tin–lead solder than epoxy polymer alone.

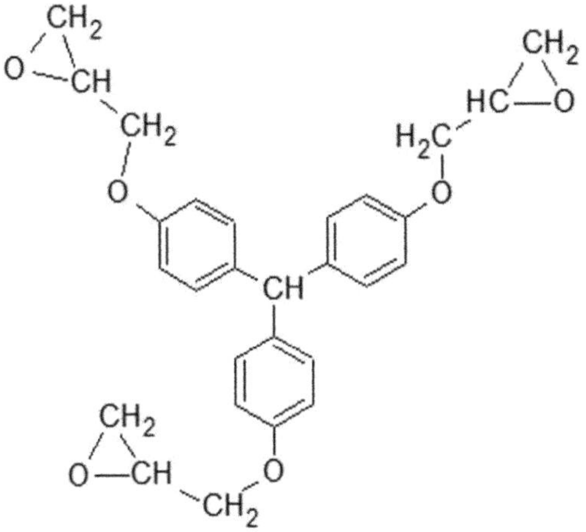

FIGURE 14.1   Structure of Huntsman Tactix 742 epoxy resin

14.7.2   *PROPERTIES OF THE SUCCESSFUL CANDIDATES*

Formulations A through D were characterized by the following properties:

1. Melting point of about 35°C.
2. A viscosity of 1000 cps at a temperature between 45 and 90°C.
3. A glass transition temperature greater than 115°C when fully cured.
4. Coefficient of thermal expansion of about 30–34 ppm/°C when fully cured.

Conversion of Formulations A through D from hot melt to cross-linked, rigid epoxy can be accomplished in less than 2 min at 150°C, or 30 min at 120°C. Processing operations, which take place after placement of the re-active hot melt encapsulant, should suffice to cure the hot melt. However, temperature rises in the application itself will further the state of cure.

14.8   SOLDER JOINT LEAD ENCAPSULATION PATENTS

Two patents were awarded covering essentially the same invention in 1998. The two patents have different titles, a somewhat different name order of the inventors, and different assignees. U.S. Patent 5,708,056 is assigned to both Delco Electronics and Hughes Electronics, whereas U.S. Patent 5,759,730 is only assigned to Delco Electronics.

14.8.1   U.S. PATENT 5,708,056

U.S. Patent 5,708,056 entitled "Hot Melt Epoxy Encapsulation Material" was awarded to Theresa Renee Lindy, Samuel R. Wennberg, Henry Morris Sanftleben, James M. Rosson, and Ralph D. Hermansen on January 13, 1998. Everyone except me was a Delco Electronics employee. The patent application was filed on December 4, 1995. The essence of the invention is that a new method of fortifying the soldered leads of surface mount integrated circuit packages. The new solder joint lead encapsulant is a reactive hot melt epoxy. Thus, lower processing temperatures can be used than the temperature required for traditional epoxy encapsulants.

14.8.2   PATENT U.S. 5,759,730

Patent U.S. 5,759,730 entitled "Solder Joint Encapsulation Material" was awarded to Ralph D. Hermansen, Theresa Renee Lindy, Samuel R. Wennberg, Henry Morris Sanftleben, and James M. Rosson on June 2, 1998. The patent application was filed on July 21, 1997. I was the only inventor from Hughes; all the others were from Delco Electronics. The essence of the invention is that a new method of fortifying the soldered leads of surface mount integrated circuit packages. The new solder joint lead encapsulant is a reactive hot melt epoxy. Thus, lower processing temperatures can be used than the temperature required for traditional epoxy encapsulants.

KEYWORDS

- **solder joint lead encapsulant**
- **epoxy resin**
- **Delco Electronics**

PART V

Custom-Formulated Organic Solder to Eliminate Lead

CHAPTER 15

FORMULATION OF A DROP-RESISTANT, ORGANIC SOLDER

CONTENTS

This project was conducted by representatives from several major companies working under the direction of the National Center for Manufacturing Science. The project began as an effort to define the property requirements needed in organic solder. The goal was to find an organic solder to replace the tin/lead solder commonly used in electronic assemblies with a lead-free alternative. All the manufacturers of organic solder were asked to submit samples and an evaluation was undertaken. None of the suppliers could pass the requirement that the organic solder pass the drop test. We at Hughes Aircraft felt that our Flexipoxy knowledge might allow us to succeed. This section is about the custom formulation of an organic solder that passed the drop test.

15.1   SCOPE OF THIS CHAPTER

There was a great deal of engineering activity by members of the NCMS team, which I am omitting from this chapter. I am narrowly focusing on one particular aspect of the formulation development, namely, the resolution of two conflicting goals. In this case, increasing the silver filler content of an electrically conducting adhesive improved the conductivity but worsened the survivability of the bonded joints to impacts from a device being dropped. Decreasing the silver filler content reversed the problem, impact resistance improved and electrical conductivity worsens. It appeared to be an irresolvable dilemma.

The NCMS effort to find and qualify an electrically conducting adhesive to replace tin/lead solder was a multi-year, highly complex project involving the some of the best engineers on the continent. My apologies to those involved, whose important contributions I have not recognized here.

15.2   WHAT IS WRONG WITH SOLDER?

Metallic solder is widely used to make interconnections in electronic assemblies. Solder is a eutectic blend of tin and lead, that is, eutectic solder melts at a single temperature. The problem is that the world is becoming increasingly aware of the toxic nature of lead. The European Union has taken a lead in banning lead from products likely to end up in a landfill. Lead in drinking water poses a serious health hazard, especially regarding brain damage to young children. American manufacturers selling to Europe have to think about alternatives to solder, which are lead free. Moreover, other countries of the world are likely to follow Europe's lead in banning lead. Lead-free alternatives are needed. Organic solder is such an alternative. Organic solder has the advantage over solder of not requiring solder flux to make an electrical connection. Solder flux must be cleaned off the PWB after soldering. Ozone-depleting Freon solvents are commonly used for this purpose.

15.3   A NCMS CONSORTIUM PURSUES AN ORGANIC LEAD-FREE SOLDER

NCMS is short for the National Center for Manufacturing Science. This organization helped underwrite projects of benefit to the nation and was participated in by consortiums of interested companies. Duane Napp was the NCMS coordinator. Deborah Huff was the original Hughes Aircraft engineer in the organic solder consortium. Robert (Bob) Doenitz took over from her in late 1994 as Hughes Aircraft engineer assigned to the consortium on organic solder.

15.3.1   CONSORTIUM MEMBERS

The members of the NCMS Alternative Interconnect Committee were as follows:

Duane Napp was the coordinator of the NCMS of the NCMS project. The participants were the companies listed in the table below: The companies were represented by the individuals also listed in the table.

TABLE 15.1   Members of the NCMS Alternative Interconnect Committee

Company	Representatives
Delphi-Packard Electric Company	Mike Zwolinski and Julie Hickman
Delco Electronics Company	Arun Chadhuri
Lucent Technologies (formerly A,T & T)	Holly Dee Rubin and Julia Zaks
Ford Motor Company	Shaun McCarthy
Celestica Company	Peter Arrowsmith
United Technologies Company	Tom Hanlon
Sandia National Laboratory	John Emerson
Hugh Aircraft Company	Deborah Huff, then Robert Dunaetz, then Ralph Hermansen

15.3.2   WHAT IS ORGANIC SOLDER?

Most of the existing organic solders are epoxy adhesives, highly filled with a silver particulate powder. Organic solders are also known as electrically conductive adhesives (ECAs). Although expensive, silver maintains conductivity even when the surfaces oxidize. In the case of other metals, like aluminum, the oxide greatly diminishes its electrical conductivity. Gold particles would work even better than silver, but gold is prohibitively expensive for the majority of applications. In order for the adhesive to be electrically conductive, there has to be a sufficiently high loading of particles so that the particles actually touch each other. At lower loadings, the non-conductive polymer coats the particles and isolates one particle from another.

Although organic solder is expensive, it solves more than just the lead problem. It also eliminates the need for flux and the chemicals used to clean up after the fluxing operation. CFCs, which are usually used to clean flux away, are damaging to the ozone layer. Eliminating the need for them is a benefit for the environment. Even so, the fact that processing steps are reduced is in itself a cost saver. Organic solder has a further advantage over tin/lead solder in that it cures at significantly lower temperatures than the melting point of solder. Lower temperature processing reduces residual stresses and increases fatigue resistance of the interconnections.

15.4   I BECAME THE HUGHES ELECTRONICS REPRESENTATIVE ON THE TEAM

I replaced Bob Dunaetz as Hughes Aircraft representative on the NCMS project in August 1995, when he was hospitalized and a few weeks prior to his death. This was my first experience of working within a consortium of major companies and it was an enriching career experience.

15.4.1   WHAT HAD PREVIOUSLY BEEN ACCOMPLISHED?

15.4.1.1   THE DEBORAH HUFF PERIOD

Deborah Huff was the first Hughes Aircraft representative on the NCMS committee. She conducted an extensive survey throughout the Hughes

Aircraft Corporation to learn as much as possible about conductive adhesives and about what parameters were important to its implementation. Her survey identified 69 different parameters. The NCMS steering committee determined which tests were most important for screening commercially available conductive adhesives, and a list of 25 conductive adhesives was established for evaluation by the NCMS group.

15.4.1.2   TWENTY-FIVE ORGANIC SOLDER CANDIDATES

Screening tests were conducted on the 25 candidates in 1994. One significant fact is that none of the organic solder candidates passed the drop test, which was devised by the committee. Many electronic devices such as phones, mobile radios, etc. get dropped occasionally. It is unacceptable that they fail so easily. The manufacturers were asked to focus on this important requirement and resubmit better candidates.

15.4.1.3   IN-HOUSE FORMULATING EFFORT

Steve Lau worked with Deborah to develop an organic solder capable of passing the drop test (ECA). They employed Flexipoxy technology and they actually succeeded. However, that candidate failed to pass a heat/humidity requirement. Deborah obtained funding for Steve to evaluate formulation variations in an attempt to improve upon the drop-resistant ECA. Due to the high cost of silver filler, non-silver fillers were used in the early trials. Initially, unfilled formulations were prepared and screened for having both high resiliency and low viscosity. Low viscosity is needed to wet and coat high filler loadings. These formulations were tested for tear resistance, hydrolytic stability, NCMS drop test, and as an adhesive, both peel strength and lap shear strength.

Enough of a formulation database was developed to move towards silver-filled trials. It was soon learned that electrical conductivity improves with higher silver loadings, whereas drop resistance worsens. Conversely, lower silver loadings improve drop resistance test results.

15.4.2 BOB DUNAETZ PERIOD

Steve Lau continued the formulation development effort under the direction of the new Hughes Representative on the committee, Bob Dunaetz. Dean Johnston is also working in the lab under Steve Lau's direction.

By February 1995, Bob reported that over 50 different formulations had been created and tested. These included both Flexipoxy and polyurethane polymers. During this period, the NCMS committee raised the drop height in the drop test to 60 inches . Five variations of best Formulation #24 were made into silver-filled adhesives and tested. Volume resistivity was under 5 milli-ohm-cm, hydrolytic stability was fair, but drop test results were only marginally acceptable. It was decided to use the Ross vacuum mixer to prepare future trials with the expectation that preparing the formulations using vacuum mixing will improve results by eliminating all entrapped air.

I was not directly involved in the formulation effort but I helped Bob do other things on the NCMS project whenever he requested my help. Bob was the department manager when I asked to transfer into TSD and he immediately approved my transfer. Over the years, Bob had funded me to do various assignments for him. I felt there was a kind of trust that existed between us.

One day a very sad and shocking event occurred. Bob was only 65, but contacted a deadly lung disease. He was hospitalized and was on a breathing machine. Due to the fact that I was familiar with the project, I temporarily took over for him as the HAC representative in the consortium. A few weeks later, Bob passed away on October 16, 1995 and I found myself with a permanent assignment.

15.5 MY FIRST MEETING WITH THE TEAM

I attended my first meeting with the working group and found myself overdressed. I was wearing a three-piece suit and tie, while all the over attendees were dressed casually. Somehow, I did not get the word about the casual attire. Being a new face in the group was drawing glances enough. Being overdressed only added to my self-consciousness. My plan going in was to sit, listen, and keep my mouth shut so that I could learn as much as possible from the conversation. However, before I knew it, I was at the white board in the front of the room informing them regarding formulating theory.

I did it because I felt that the group was lost in a hopeless quest. Their problem was that they had identified about 30 different property requirements for their ideal organic solder. My experience had taught me that they had defined an impossible task. Using set theory Venn diagrams, I illustrated why I thought they needed to reduce the list of requirements drastically. I suggested that they select the five most important properties and focus on them in the short run.

The committee did just that. Table 15.2 shows the five most important properties which we would be focusing on.

TABLE 15.2   Most Important Property Requirements for NCMS Organic Solder

Property	Requirement	Comment
Drop resistance	6 drops minimum	See Note 1
Volume resistivity	Less than 1×10^{-3} ohm-cm	See Note 2
Cure schedule	Less than 30 min at 150°C	
Work-life	More than 4 h at 25°C	
Aging at 85°C/85% RH	Less than 20% change in resistance	

Note 1: Conductive adhesive layer is 5–7 mils. Test Specimen is a 44 pin, leaded PLCC bonded to an epoxy-glass rectangular board. Three specimens are individually dropped from 60 inch height vertically down a guide so that they hit on their edge. The number of drops survived is recorded.

Note 2: Adhesive layer is 5–7 mils. Test component consists of a polyimide coupon with four point test traces of OSP copper and Sn–Pb.

15.6   SHOULD THE FORMULATING EFFORT BE HALTED?

My management was concerned that the formulating effort had gone on for well over a year without resolving the problems. The formulation effort had been underway for years and countless formulation variations had already been tried. The effort started under Deborah Huff in 1994, was continued under Robert Dunaetz in 1995, and now I was continuing it in Fall 1995. As I came fresh into the job, they instructed me to determine if we had a reasonable chance of ever meeting the NCMS requirements or if we should abort the formulation effort and use the precious funding elsewhere. The appearance was that the requirements of drop test survival and good electrical conductivity (i.e., low volume resistivity) were in direct opposition to each other and may never be resolved. Higher silver filler loadings improved conductivity but diminished drop resistance. Lower silver filler loadings improved drop resistance but diminished the conductivity.

The following table shows the dilemma. Flexipoxy 12 was prepared at various silver filler levels and tested for volume resistivity and drop resistance. It takes more than 80% silver to pass the less than 1 milli-ohm cm volume resistivity requirement, but less than 81 % silver to pass the drop test requirement of six drops minimum. There seems to be no zone of silver level where both requirements are met.

TABLE 15.3 DuPont V-9 Silver Filler Level Effect on Volume Resistivity and Drop Resistance

Property	Flex. 12-79	Flex. 12-80	Flex. 12-81	Flex. 12-83	Flex. 12-85
Wt.% of DuPont V-9 Silver powder	79	80	81	83	85
Volume resistivity, milli-ohm-cm	5.62, Fails	1.42, Fails	0.64, Passes	0.32, Passes	0.16, Passes
Average number of drops	12, Passes	6, Passes	5, Fails	3, Fails	2, Fails

The next experimental series would decide that question. Silver fillers from several sources had been ordered. These different fillers were each incorporated into our best Flexipoxy formula and the adhesives tested for volume resistivity and drop resistance survivability.

We are looking back in time so we have the answer. One of the 10 silver fillers did prove successful. Were we lucky? Yes, but that luck happened due to perseverance and determination. I was able to enthusiastically recommend to management that we complete the formulation work and vigorously pursue a patent. In the next section, I will describe the developmental process to you.

15.7 THE SILVER FILLER SELECTION EXPERIMENT

15.7.1 THE BASE FORMULATION

Flexipoxy-12 ECA, which contained DuPont K003L silver filler, had variable results in that it had both passed and failed the conductivity and drop test requirements. We decided to use it as the basic formulation for a comparison of 10 different silver fillers. Each of the trial fillers would replace the K003L silver filler in the formulation at an 80% loading level. Volume resistivity of each filler would then be determined. The formulation is presented in Table 15.4.

TABLE 15.4   The Formulation for Flexipoxy-12 ECA

Ingredient	Weight percent
Epon 825 Rigid Epoxy Resin	5.9
Heloxy 67 flexible epoxy resin	4.7
DER 732 flexible epoxy resin	1.2
Humko DP-3680 flexible curative	4.1
ATBN flexible curative	4.1
Mallinkrodt BYK-052 wetting agent	0.1
Kenrich KR-9S titanate coupling agent	0.1
DuPont K003L silver filler	79.9

15.7.2   THE SILVER PARTICULATE FILLERS TO BE EVALUATED

The following table shows the different silver fillers which were evaluated and whether they were electrically compliant with the requirement that volume resistivity be 1 milli-ohm cm or less.

TABLE 15.5   The Silver Fillers Evaluated and Their Volume Resistivity Results

Supplier	Filler designation	Electrically compliant
DuPont	K003L	None
Potters	SC500P18, SC140F19U, SH400S33, SM325F55, SM140F65	None
Degussa	9al, 26lv, 52, 80,95	None
Technic	Silflake 237, 282, 299, 450, 499	299, 450

15.7.3   TEST RESULTS

The list of winners was small.

1. Technic 299 and 450 produced the lowest volume resistivities, markedly lower than any of the Potters or Degussa silver fillers. The Technic 299-filled adhesive had a volume resistivity of 0.59 $\times$ 10^{-3} ohm-cm and the Technic 450-filled adhesive had a volume resistivity of 0.16×10^{-3} ohm-cm.
2. The DuPont K003L-filled adhesive was third best with a volume resistivity of 1.42×10^{-3} ohm-cm.

We decided to focus on Technic 299 and 450.

15.8   THE SILVER LOADING LEVEL EXPERIMENT

15.8.1   PREPARING MASTERBATCHES

Now, we needed to understand how silver filler content of these superior fillers affected drop resistance and conductivity. The new experiment again used Flexipoxy-12 as the base formulation. Masterbatches were prepared containing two different silver fillers with the intent of blending them to attain intermediate filler levels. Table 15.6 shows the designation, silver filler, and weight percent of each.

TABLE 15.6   Masterbatches **for the Fine-Tuning Silver Filler Experiment**

Designation	Silver filler	Weight percent
Flexipoxy-450-70	Technic Silflake 450	70
Flexipoxy -450-80	Technic Silflake 450	80
Flexipoxy -299-70	Technic Silflake 299	70
Flexipoxy -299-80	Technic Silflake 299	80

15.8.2   PREPARING TRIALS AT SPECIFIC SILVER FILLER LEVELS

Next, the Technic 450 masterbatches (Flexipoxy-450-70 and Flexipoxy-450-80) were blended to produce intermediate loading levels and the Technic 299 masterbatches (Flexipoxy-299-70 and Flexipoxy-299-80) were also blended to produce intermediate loading levels.

15.8.3 THE TEST RESULTS

The intermediate trials were representative of 76, 78, 79, and 80 weight percent silver filler. These conductive adhesives candidates were tested for drop resistance and volume resistivity. A summary of the results is shown in the Table 15.7.

TABLE 15.7 Summary of Filler Level Optimization Results

Silver level, Wt. %	Silver filler	72	76	78	79	80
Volume resistivity, milli-ohm cm.	299	No test	0.96	0.66	0.63	0.46
Volume resistivity, milli-ohm cm.	450	1.17	0.27	0.15	No test	0.10
Number of drops*	299	No test	14.8	8.2	14.2	10.2
Number of drops*	450	6.0	3.1	2.0	No test	2.7

*Average of three specimens.

Comments:

(1) Technic 299 was the only silver filler able to attain both acceptable volume resistivity and acceptable drop test results in a single formulation. Not only was it able to satisfy both requirements, but it did it over a range of filler loadings, namely, 76, 78, and 80%. It is possible to formulate at 78% and have a cushion of safety around this level. All future development will be based on this filler (Technic 299).

(2) Technic 450 filler was the second best filler of all examined. Unfortunately, there was no filler level at which it passed both conductivity and drop resistance requirements.

15.9 THE CURATIVE BLENDING EXPERIMENT

15.9.1 IMPROVING DROP RESISTANCE THROUGH CURATIVE BLENDING

ATBN 1300X16 is often used in adhesive formulations to enhance toughness and reduce brittleness of the adhesive. The rubber segments in the

ATBN are said to coalesce and form discrete rubber micro-balls within the rigid epoxy matrix, thus providing crack stopping entities. The goal of this experiment was to improve drop resistance of the ECA.

ATBN has a molecular weight of 3800 and it reacts with terminal epoxy groups via its terminal amine groups. In between them is a copolymer chain of butadiene-acrylonitrile where the acrylonitrile segment is 18%. In the study below, the ATBN content in the curative blend is varied in order to determine its importance to the key properties of electrical conductivity and drop resistance.

15.9.1.1 FLEXIPOXY 7400, THE BASE FORMULATION

A series of formulation optimization experiments was begun in late 1996, all based on the NCMS-7400-78 formulation, which is shown in the Table 15.8.

TABLE 15.8   Flexipoxy-7400-78 Formulation

Source	Ingredient	Weight percent
Shell Chemical	Epon 825	4.4
Shell Chemical	Heloxy 67	3.3
Dow Chemical	DER 732	3.3
Humko	DP 3680	11.0
Kenrich	KR-9s titanate coupling agent	Trace
Technic	Silflake 299 silver flake	78.0

15.9.2 BLENDING IN ATBN CURATIVE STEPWISE

The ATBN content in the total curative blend was varied as follows: 0, 15, 30, 45, and 60%. Drop test samples were fabricated with bondline thickness varying (5, 6, and 7 mils thickness). Silver content was held constant at 78%. The Flexipoxy designation in this experiment identifies the composition. So the name 7415-78 would have 15% ATBN in the amine blend and have 78% silver filler level.

15.9.3   TEST RESULTS

Table 15.9 shows the test results from the curative blending experiment. Volume resistivity is reported for both copper traces and for tin/lead traces. Drop test results are highly dependent on bondline thickness and results are shown separately for 5, 6, and 7 mil bondlines. The individual number of drops for each specimen was shown. The pass/fail criteria are that the average of three specimens must equal or exceed six drops.

TABLE 15.9   Volume Resistivity and Drop Resistance for the Curative Blending Experiment

Property	Flexipoxy 7400-78	Flexipoxy 7415-78	Flexipoxy 7430-78	Flexipoxy 7445-78	Flexipoxy 7460-78
Volume resistivity, milli-ohm-cm	1.07 (Cu) 1.02 (Sn/Pb)	0.90 (Cu) 0.87 (Sn/Pb)	0.50 (Cu) 0.46 (Sn/Pb)	0.65 (Cu) 0.57 (Sn/Pb)	13.2 (Cu) 7.2 (Sn/Pb)
Number of drops, 5 mil bondline	4, 5, and 7, failed	4, 5, and 8, failed	3, 4, and 7, failed	2, 2, and 4, failed	2, 2, and 3, failed
Number of drops, 6 mil bondline	5, 6, and 7, passed	5, 6, and 8, passed	11, 11, and 15, passed	5, 8, and 12, passed	5, 6, and 9, passed
Number of drops, 7 mil bondline	6, 7, and 9, passed	12, 14, and 16, passed	18, 19, and 23, passed	8, 12, and 22, passed	8, 9, and 16, passed

15.9.4   THE DISCRETE MICRO-BALL THEORY

The first surprise was how volume resistivity was affected. Flexipoxy 7400-78 (i.e., 0% ATBN)was a tad above 1 milli ohm-cm. 7415-78, 7430-78, and 7445-78 were all well below 1 milli ohm-cm but 7430 was the best candidate. 7460-78 was not electrically conductive even though it had the same silver content as the other ones, which passed. One possible explanation for this strange behavior is that at the highest ATBN level, the ATBN tends to form large enough micro-balls of rubber in the epoxy matrix that the balls isolate silver particles from each other.

15.9.5 FORMULATION 7430-78 IS BEST

Formulation 7430-78 had best drop resistance of all the variations. I say this ignoring the 5 mil bondline sample which failed. However, none of the variations in this experiment passed at 5 mil bondline thickness. Table 15.10 shows the formulation.

TABLE 15.10   Flexipoxy 7430-78 Formulation

Source	Ingredient	Weight percent
Shell Chemical	Epon 825	4.0
Shell Chemical	Heloxy 67	3.0
Dow Chemical	DER 732	3.0
Humko	DP 3680	8.3
Hycar (now Hypro)	ATBN 1300X16	3.6
Kenrich	KR-9s titanate coupling agent	Trace
Technic	Silflake 299 silver flake	78.0

15.10   FLEXIPOXY 7415 SILVER CONTENT VARIATION STUDY

15.10.1 EXPERIMENTAL

So far we have made good progress in attaining both good electrical conductivity and good drop resistance in a single formulation. Now, we attempted to find the exact amount of silver filler to optimize these two properties. To this end, an experiment using Flexipoxy 7415 base formulation was conducted, where the Silflake 299 silver filler content was varied from 76 to 82% and the bondline was evaluated at 5, 6, and 7 mils thickness.

15.10.2 TEST RESULTS

The following table shows the volume resistivity and drop test results for the silver filler content study:

TABLE 15.11   Volume Resistivity and Drop Resistance for Specific Silver Filler Experiment

Property	76 % loading	78 % loading	80 % loading	82% loading
Trial name	7415-76	7415-78	7415-80	7415-82
Volume resistivity, nilli-Ohm-cm	2.43 (Cu) 1.72 (Sn/Pb)	0.90 (Cu) 0.76 (Sn/Pb)	0.92 (Cu) 0.62 (Sn/Pb)	0.48 (Cu) 0.41 (Sn/Pb)
Number of drops, 5 mil bondline	4, 6, and 11, passed	4, 5, and 8, failed	7, 7, and 14, passed	3, 3, and 3, failed
Number of drops, 6 mil bondline	8, 16, and 19, passed	5, 6, and 8, passed	6, 8, and 8, passed	4, 5, and 6, failed
Number of drops, 7 mil bondline	10, 24, and 24, passed	9, 11, and 14, passed	9, 11, and 15, passed	7, 7, and 10, passed

Comments:

The most obvious conclusion was that adequate bondline thickness is highly important to attaining acceptable drop resistance values. The thicker the bondline, the better the drop test results. At 7 mils thickness, every trial easily passed. However, at 6 mils thickness, 7415-82 failed, 7415-80, and 7415-78 marginally passed, and 7415-76 easily passed. At 5 mils thickness, only 7415-80 passed.

All the variations except 7415-76 passed the 1 milli ohm-cm requirement for volume resistivity.

15.11   FLEXIPOXY 7415 SILVER CONTENT VARIATION STUDY

A similar experiment using Flexipoxy 7430 base formulation was conducted. The silver filler content varied from 76 to 82% and the bondline was evaluated at 5, 6, and 7 mils thickness.

FINDINGS:

Again, the most obvious conclusion is that adequate bondline thickness is highly important to attain acceptable drop resistance values. At 7 mils thickness, every 7430 variation easily passed. However, at 6 mils thickness, higher loaded trials 7430-80 and 7430-82 failed, whereas 7430-80 and 7430-78 very easily passed. At 5 mils thickness, only 7430-76 was close to being acceptable. All the variations, except 7430-76, passed the 1 milli ohm-cm requirement for volume resistivity. The NCMS 7430-78 formulation had the best all round properties.

15.12   FLEXIPOXY 7430-79 EVALUATION

We are at the end of the formulation optimization process; next we will characterize the finalist. Flexipoxy 7430 was prepared at 79% silver content and subjected to additional testing. The following table shows the results:

TABLE 15.12   Properties of Flexipoxy 7430 at 79% Silver Filler Content

Property	Result	Comment
Brookfield viscosity at 23°C		
@ 1 rpm	325,000 cps	Typical non-Newtonian pseudoplas-
@ 5 rpm	104,000 cps	tic behavior for a highly filled liquid
@ 10 rpm	68,000 cps	
@20 rpm	46,000 cps	
Work-life, h	10	Based on viscosity vs. time
		graph
Glass transition temperature	−15°C	
Thermal expansion		
Below T_g	55 ppm/°C	
Above T_g	140 ppm/°C	
Volume resistivity	0.0005 ohm-cm	Excellent
Drop resistance	15.3 drops avg.	Excellent

15.13   THE FINAL FORMULATION FLEXIPOXY 7430

15.13.1   THE FINALIST FORMULATION

The table below shows the final formulation to conclude the development work in this chapter:

TABLE 15.13   Flexipoxy-7430 Formulation

Source	Ingredient	Weight Percent
Shell Chemical	Epon 825	3.8
Shell Chemical	Heloxy 67	2.9
Dow Chemical	DER 732	2.9
Humko	DP 3680	8.0
Hycar (now Hypro)	ATBN 1300X16	3.4
Kenrich	KR-9s titanate coupling agent	Trace
Technic	Silflake 299 silver flake	79.0

15.13.2   FLEXIPOXY 7430 MANUFACTURING PROCESS

1. Weigh 36 g of ATBN 1300X16 plus 84 grams of DP-3680 into a polypropylene beaker and blend them together using a stainless steel spatula.
2. Weigh 455 g of SilFlake 299 silver flake into a separate polypropylene beaker.
3. Combine the blended hardeners and the silver flake using a stainless steel spatula thoroughly until silver flake is wetted.
4. Set the gap on the 3 roll mill to 1–2 mils. Pass the mixture from step 3 through the mill four times. Store the milled mixture in a covered polypropylene beaker until needed.
5. Add 540 g of milled silver paste from step 3 to the Ross one quart vacuum mixer. Seal the mixer, draw a vacuum, and set the planetary mixing head to turn at 33 rpm. Mix for 20 min under a vacuum pressure of 300 microns or less.
6. Weigh the following ingredients into a polypropylene beaker: 40 g of Epon 825, 30 g of DER 732, 30 g of Heloxy 67, and 0.5 g of KR-9S titanate. Blend the ingredients using a stainless steel spatula.
7. Weigh out 375 g of SilFlake 299 silver flake into a separate polypropylene beaker. Blend the silver flake into the resin blend of step 6 using a stainless steel spatula. Mix thoroughly until silver flake is wetted.
8. Set the gap on the 3 roll mill to 1–2 mils. Pass the resin/silver mixture from step 7 through the mill four times. Store the milled mixture in a covered polypropylene beaker until needed.

9. Transfer 450 g of the resin/silver mixture from step 8 to harder/silver mixture already in the Ross mixer. Seal the mixer, draw a vacuum, and set the planetary mixing head to turn at 33 rpm. Mix for 20 min under a vacuum pressure of 300 microns or less. NOTE: It is important to freeze the product from step 9 as quickly as possible!
10. Transfer the product from step 9 into pre-labeled Semco cartridges. Avoid introducing air bubbles into the product. Submerge the filled Semco cartridges in liquid nitrogen. Store the cartridges at −40°C.
11. Fill smaller cartridges (e.g., 2 cc syringes) using the thawed Semco cartridges from step 10. Flash freeze the smaller cartridges in liquid nitrogen and store them at −40°C.

15.13.3 PROPERTIES OF THE FINALIST

Table 15.14 summarizes the key properties of Flexipoxy 7430 organic solder.

TABLE 15.14   Properties of Flexipoxy 7430 Conductive Adhesive

Property	Value
Freezer storage life	3 months minimum at −40°F
Brookfield viscosity	
At 5 rpm:	104,000 cps
At 20 rpm:	46,000 cps
Work-life at 25°C (75°F)	10 h minimum
Cure schedule	
(a) For fast production	a) 10 min at 150°C
(b) For immediately properties	b) 20 min at 150°C
Drop resistance	15.3 drops avg.
Volume resistivity	0.5×10^{-3} ohm-cm
Glass transition temperature	-15°C
Coefficient of thermal expansion	
a) Below T_g	a) 55 ppm/°C
b) Above T_g	b) 140 ppm/°C

15.14   POSTSCRIPT

The story can end here in terms of formulation development. However, Steve Lau and I had a lot to do before our involvement ended. The following post activities occurred:

15.14.1   A ROUND ROBIN DROP TEST PROJECT

When the committee members attempted to verify our results using three of the members to conduct their own testing, lower results were obtained for drop resistance than Steve Lau had found at our test site. In order to help find the source of the variance, a Round Robin test plan was set up where each of the four sites (Delphi-Packard-Electric, Delco Electronics, Ford Motor Company, and Hughes Aircraft) made drop test specimens and distributed them to the other sites. Every site ended up testing a specimen set from each of the four sites. Again, bondline thickness was a major variable. Improvements to the test method resulted in more consistent results.

15.14.2   SELECT A COMMERCIAL MANUFACTURER TO MAKE FLEXIPOXY 7430 ECA

Part of our contractual obligation with NCMS was to select a qualified commercial adhesive manufacturer and to transfer the technology for making Flexipoxy 7430 to them. Many companies expressed interest in doing this. We selected the best five companies and then thoroughly screened them. Dexter Electronics Materials Division in City of Industry, California was selected and the technology transfer went very smoothly. They immediately took an active role in working with the committee members, presenting an introductory vugraph presentation to the NCMS committee at a meeting in Kokomo, Indiana and soon providing samples of Flexipoxy 7430 to them.

15.14.3   NCMS MEETING AT OUR FACILITY

During February 29 through March 1, 1996, Hughes hosted the NCMS meeting in El Segundo, California. The major presentation was given by

Steve Lau, where he recapped the history of the formulation development effort. Flexipoxy 7340 was the end product of that extensive project. I will never forget that time because amidst the fervent activity, I was processing out as a Hughes employee due to my retirement. I immediately returned to work as a contract engineer the next day because I had too many project commitments at that time to just walk away from them.

15.14.4   PREPARE AND ISSUE A FINAL REPORT TO NCMS

A 55 page final report titled, "Final Report on the Development of Lead-Free Organic Solder" (Hughes Ref. No. E6062) was submitted to NCMS in November 1996. The authors were Steve Lau and Ralph Hermansen.

15.15   PATENT

15.15.1   U.S. PATENT 5,929,141

U.S. Patent 5, 929,141 entitled, "Adhesive of Epoxy Resin, Amine-Terminated BAN, and Conductive Filler" was awarded to the inventors: Lau, Steven E., Huff, Deborah S., Hermansen, Ralph D., and Johnston, E. Dean on July 27, 1999. The formulation development work was done by Hughes Aircraft Company and the patent application was filed on July 28, 1997. However, between the date of filing and award, Raytheon Company acquired most of the Hughes Aircraft Company. When the patent was issued in July 1999, it was assigned to the Raytheon Company.

15.15.2   THE TITLE OF THE PATENT COULD HAVE BEEN BETTER

I received a copy of this patent after I had fully retired and immediately wondered what the word "BAN" in the title meant. It is short for butadiene-acrylonitrile, which is a block terpolymer. I'll bet others are equally bewildered. BAN is the polymeric part of one of the flexible curatives in the formulation. I had earlier told our patent attorneys that the patent should have been titled "Drop Resistant, Electrically-Conductive Adhesive." The big breakthrough in this invention is the attainment of both good drop resistance and good electrical conductivity in one adhesive. That was really the major accomplishment of this formulating effort. I was no longer a "true" Hughes employee at the time of filing. I was only a contract engineer, working beyond my retirement from Hughes Aircraft Company. I guess they felt free to ignore my requested name for the patent.

15.15.3 ABOUT THE CO-INVENTORS

There are four inventors on this patent and it surely would have also included Bob Dunaetz, had he lived. Steve Lau leads the list and that honored position is deserved. He was the formulating innovator throughout the project. Deborah Huff did the pioneering work as the first Hughes Electronics representative. She was freed up from this project so she could focus another more pressing assignment. The original invention disclosure began with Deborah Huff and was updated over time by Steve Lau. I strongly agree that Steve Lau deserves having first position in the inventor list. He worked for nearly 3 years on this effort and was the source of the formulation planning for most of it. E. Dean Johnston worked on formulation preparation and testing with Steve Lau in the lab.

KEYWORDS

- **organic solder**
- **NCMS**
- **Hughes Aircraft**

PART VI
Rigid Polyurethanes for Aircraft Transparencies

CHAPTER 16

SIERRACLAD™ BIRD-PROOF CANOPIES AND WINDSHIELDS

CONTENTS

This was an important and very challenging, custom-formulating project. It was important because our country's F-15 fleet was handicapped by its inadequate canopies. They couldn't fly in the rain for fear of losing the protective coating on the polycarbonate surface. This project had the potential of resolving that problem. The project was challenging because the demands on the F-15 canopy are extreme. This plane flew at high speeds at a very low elevation to avoid enemy radar. The likelihood of hitting a large bird was high enough that a very impact resistant construction was crucial.

From a formulator's viewpoint, this project was fascinating. Principles of polymer science were put into practice on a daily basis. For example, I was able to examine the effects of cross-linking on the properties of the rigid transparent polyurethane plastics, which I was formulating. Some of my candidates were equal to polycarbonate in impact resistance, and I think that is an impressive thing. The whole technology, which I developed, might have been lost to the world without this publication. Fortunately for us, when Sierracin patented this technology, they placed it in the public domain and I am able to tell its story.

16.1   SCOPE

The engineering project discussed here involves the development of protective rigid polyurethane cladding for F-15 polycarbonate canopies. A team of engineers worked on this project for many months. Some of them specialized in forming canopies and windshields with the new materials. My involvement was mainly directed to formulation refinement to meet the many property requirements on the material.

However, there is another story that is concurrently told and it is the story of a whole new genre of plastics being developed. There is a niche area of polyurethane technology which is so rare that it is practically unknown. That is the area of thermosetting rigid polyurethane plastics. There was no conscious intent to develop them, but in the process of attempting to meet the many property demands of the F-15 canopy problem, they were created. Numerous formulations were developed to attain certain properties, for example, hardness, impact resistance, abrasion resistance, transparency, stretchability during forming, thermal behavior, and outdoor weathering. Some formula variations were as impact resistant as polycarbonate itself. Other variations could deform enough to form deep domes. Others were nearly as hard and brittle as epoxy plastics.

16.2 BACKGROUND

16.2.1 THE F-15 AIRCRAFT CANOPY MUST BE IMPACT RESISTANT

The F-15 airplane was designed to fly very fast and low to the ground in order to avoid enemy radar detection. The original F-15 canopy was formed from acrylic plastic, but it was soon discovered that acrylic canopies were far too fragile for this application. If the pilot accidentally flew into a large bird, such as a goose, the acrylic canopy would shatter with dire consequences for the occupants and for the plane itself. Fortunately, there was an alternate transparent plastic having far better impact resistance. Thus, the next canopies were formed from polycarbonate plastics. Part of the military canopy evaluation process is to subject test canopies to a bird impact test. As expected, the polycarbonate canopies were totally acceptable from a bird-impact standpoint, but they had other problems. Unlike acrylic plastics, polycarbonate is too soft to be optically polished. The mars and scratches from normal use soon caused the polycarbonate canopies to lose their transparency. Moreover, polycarbonate is adversely affected by certain chemicals. Crazing and loss of transparency can result from exposure to the more aggressive of these chemicals. Polycarbonate surfaces must be protected from these agents.

For those unfamiliar with the term "crazing," it is a phenomenon which occurs in polymers, where narrow microscopic gaps develop in the material. Crazes are only visible because light reflects off the surfaces of the gaps. A craze is different from a crack because it cannot be felt at the surface and the plastic can still support a load. Crazes form at stressed areas and tend to propagate perpendicular to the applied tension. Polymers such as polystyrene, acrylics, and polycarbonate are especially prone to crazing.

16.2.2 THE FUSION-BONDED ACRYLIC/POLYCARBONATE PROBLEM

David Voss was my manager at Sierracin-Sylmar. However, before that he had worked at McDonald-Douglas in Saint Louis, during this F-15 canopy design period. Well aware of the F-15 canopy problem, he had a patentable idea: Laminate thin acrylic facings to the polycarbonate main ply and the outer surfaces of the canopy would be harder and capable of optical rejuvenation via polishing techniques. His selected process of uniting the two plastics was fusion bonding. At elevated temperatures and pressures, the polymer chains of the two plastics diffuse into one another creating a strong bond between them. Dave expected to enjoy the best of both worlds. He would retain the superb impact properties of polycarbonate, while having the superior abrasion resistance and chemical resistance of acrylic plastics. So, the problem encountered was unexpected. The impact resistance of the fusion-bonded composite was far less than that of polycarbonate and much closer to that of acrylic plastic.

16.2.3   AN ULTRATHIN HARD-COAT SOLUTION

When I joined Sierracin/Sylmar in 1974, David Voss had already left McDonald-Douglas to join Sierracin, where he headed up the M and P Engineering department. Bill Miller was my immediate supervisor. Sierracin was producing polycarbonate canopies for the F-15 aircraft and they had developed a special process for protecting their surfaces from an abrasion and hostile chemicals. They applied a propriety thin coating to the unformed, flat polycarbonate sheets, which hardened the surfaces to abrasion and protected the polycarbonate from hostile chemicals. The coating was thin enough to have no negative effect on the impact resistance of the formed canopies. It appeared to be a utopian solution except for one problem. If the F-15 aircraft flew through clouds at high altitude, ice crystal abrasion removed the coating and the bared polycarbonate surface became marred. In other words, the U.S. Air Force had a fleet of F-15 airplanes which they dare not fly in high clouds.

16.3   THE PROBLEM

The Air Force problem is that the fleet of F-15 aircraft could not fly in the rain for fear of losing the protective coating on the polycarbonate surface. For Sierracin Company, this problem was an opportunity in that the supplier first able to build a suitable F-15 canopy is likely to own that business for a long time. For the Engineering Department at Sierracin, the problem was to win that race between windshield/canopy suppliers by solving the technical problem. I was part of that department and the originator of the first transparent rigid polyurethane plastic samples. My role in the project was to modify the formulations as needed to satisfy the objectives.

16.4 A NEW POTENTIAL SOLUTION

My new colleague, John Raffo, and I soon became friends and sometimes stayed after normal work hours when we were free to do our own experiments without requiring anyone's approval. John had an aptitude and interest in instrumental chemical analysis. During our conversions, I mentioned that non-yellowing, water-white, rigid polyurethanes might be an interesting transparent material. He ordered the ingredients and when they arrived, we made castings and examined them.

16.4.1 SERENDIPITY

One of those evenings, Dave Voss wandered into the lab and wanted to know what we were doing and who had authorized it. We told him it was a product of our own initiative and we did it on our own time. Dave turned the polyurethane (PU) transparency over and over as he pondered something. He said let's see if it will adhere to polycarbonate and what the impact resistance of the composite part will be. The hope was that polyurethane would not initiate a crack in the polycarbonate, as the fusion-bonded acrylic did.

Preliminary testing of the water-white rigid, polyurethanes looked favorable. They could be cast against a sheet of polycarbonate and adhered tenaciously. The impact resistance of the composite was still better than that of acrylic. So Dave pitched the idea to upper management and returned with a directive to pursue the idea on a priority basis.

16.4.2 WATER-WHITE, NON-YELLOWING, RIGID POLYURETHANES

Table 16.1 shows the formulations of the early rigid polyurethanes that John and I formulated in our free time. Pluracol TP-440 and tetrol PEP 550 have polyoxypropylene chain segments in them. Teracol 1000 is a diol having a polyoxybutylene chain segment. Hylene W is an aliphatic diisocyanate. Specifically, it is 4,4'-methylene-bis-(cyclohexyl isocyanate). A few drops of di-butyl tin di-laurate (DBTDL) were used to catalyze the polymerization reaction. The names Clad 4, Clad 9, Clad 8, and Clad 5 were assigned later when the Sierraclad™ concept was formalized.

TABLE 16.1 Formulations of the Early Clad Materials

Ingredient	Clad 4	Clad 9	Clad 8	Clad 5
Pluracol TP-440 triol	51.58%	46.42%	41.26%	
Teracol 1000 diol		7.83%	15.67%	
Pluracol PEP 550 tetrol				47.53%
Hylene W diisocyanate	48.42%	45.75%	43.07%	52.47%

The chemical structure of Hylene W is shown below:

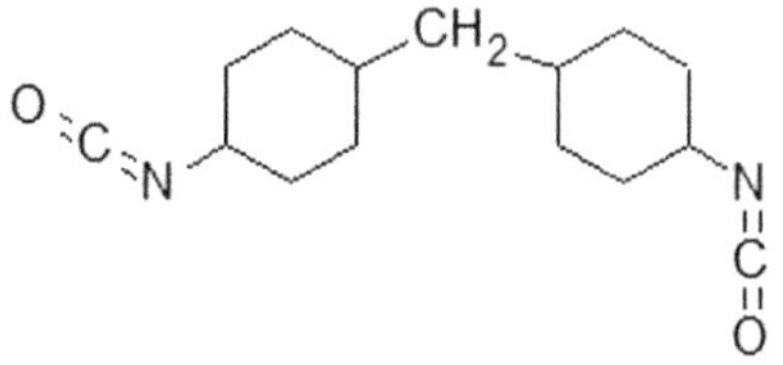

FIGURE 16.1 Chemical structure of Hylene W diisocyanate.

It was important to process the polyols to remove all absorbed water, which would cause foaming. Thus, the polyols were heated to 220°F with stirring and a vacuum applied to remove the water vapor. Thereafter, the dried polyols were stored in metal cans with screw type lids.

16.5 THE SIERRACLAD™ PROJECT

A trade name for the concept of polyurethane-clad polycarbonate was quickly conceived. The concept was called Sierraclad. Before long, John Raffo and I were attempting to produce large sheets of the Sierraclad composite to demonstrate scale-up feasibility. In order to obtain smooth optical surfaces, we cast the polyurethane in molds utilizing tempered glass buffers. Most of these casting attempts failed due to the polyurethane's ability to defeat all of the different mold release agents which we tried.

The polyurethane adhered strongly to the glass surfaces and pulled glass chips.

16.5.1 MY FRUSTRATION WITH THE SIERRACIN DEVELOPMENTAL APPROACH

If I had been allowed to run this development effort my own way, I would have spent the next few months evaluating different formulations and processes at perhaps a one-twentieth scale of the actual parts. I think my approach would have been more economical and certainly less arduous. However, companies have personalities reflective of their founders. The management wanted to see very rapid scale-up. John and I spent the following months trying to cast large specimens against glass buffers to attain an optical surface on the PU facings. It was a frustrating period for us. Our problem was finding a 100% effective release agent for casting it against glass surfaces. The polyurethane would adhere strongly to the glass even if there was only a pinpoint area free of release agent. Then a glass chip would be embedded in the PU plastics surface. We ended up destroying a dozen or more of the large glass buffers due to the mold release problem.

In fairness to Sierracin Management, their approach was justified by the urgency of the Air Force problem and the value of spotting scale-up problems as early as possible.

16.5.2 A WORK-AROUND TEMPORARY SOLUTION

Finally, Dave Voss told us to regard the mold release problem separately and to continue to make PU-clad polycarbonate specimens by covering the glass buffers with polypropylene film pulled taut by heat shrinking on to the glass. This change ended the glass breakage problem and allowed us to make and test specimens for impact resistance and for forming the clad sheets into domes. Some specimens were sent out for outdoor weathering exposure. Some were tested on Sierracin's own weatherometer. The polypropylene film was a stopgap solution which allowed us to proceed with the other testing, but was not an acceptable method for producing optically acceptable, cast surfaces on the rigid polyurethane.

Despite the setbacks, the company was moving forward with the assumption that the concept would be eventually successful. Dave had a patent application prepared and the sales force was out talking our new invention up to our customers. I learned from talking with our sales engineers that "Sierraclad" was all the buzz in Air Force circles.

16.5.3   THE NEW POLYURETHANE PLASTICS WERE MY RESPONSIBILITY ALONE

All of the discoveries related to the cladding of polycarbonate with polyurethane transparent plastics were clearly the property of Sierracin Corporation. However, the fact that a patent was granted covering that technology and has since expired, puts the information in the public domain. However, when it comes to formulation development, no one at Sierracin ever gave me advice on how to formulate rigid polyurethanes nor was anyone there knowledgeable enough to advise me. The opposite is true. I taught the people assisting me how to process the ingredients and so on. The capability of formulating polyurethanes was my sole province, developed from my previous experience and education. Therefore, I regard the modification and application of the formulations to new uses as originally belonging to me.

16.6   WHAT PROPERTIES MUST THE CLADDING HAVE?

Although no one formally drew up a list of properties that the cladding must process at the time, my managers had sufficient experience with aircraft windshield and canopy engineering to compile it if they had so wished. I was new to the transparency field and learned on the job what was needed in the cladding material. One of the requirements for the polyurethane cladding is to survive the forming operation at the forming temperature of polycarbonate (i.e., from 270 to 375°F). The cladding must remain adhered to the polycarbonate and not rupture due to the stretching process. There were many other requirements in addition to the forming requirement.

Looking back in time, I would now construct a table of required properties as follows:

TABLE 16.2 Expected Cladding Properties and Simulated Testing

Canopy requirement	Simulated testing	Expectations
Transparency over time:		
1. Low color	Expose to sunlight	Minimum change
2. Low haze	Taber Abraser plus haze reading Salt blast abrasion	Change in haze is key property
Bird strike protection	Dart impact test	As close to polycarbonate impact as possible
Resistance to normal chemicals	Expose cladding to isopropanol, methyl ethyl ketone, etc.	No crazing or damage Minimum change
Formability		
1. Stretch	Dome thermo-forming	Maximize depth of dome formed
2. Adhesion to polycarbonate	Thermal cycling	No delamination
Temperature service range		
(−60°F to 160°F)	Test at high and low temperatures	Preservation of required properties
1. Durometer		
2. Tensile properties		
Weathering	Intense UV exposure	Minimal changes
	Heat plus humidity aging	
	Weatherometer	
	Outdoor exposure	

16.7 PRESENTATION OF THE EXPERIMENTAL TASKS

The rest of this chapter consists of numerous experiments conducted in order to custom formulate a transparent rigid polyurethane plastic capable of meeting all of the objectives described above. I decided to present these data under each of the properties sought as follows:

1. Low haze and high transparency
2. Protection from normal chemicals
3. Temperature service range
4. Bird strike protection
5. Canopy formability
6. Weatherability

16.8   LOW HAZE AND HIGH TRANSPARENCY

16.8.1   NON-YELLOWING CHARACTERISTIC

The transparent, rigid, non-yellowing, polyurethane plastics that I had been examining were water-white and highly transparent. The polyurethanes were formulated using only aliphatic isocyanates and that is the secret to preventing yellowing from sunlight. Most polyurethanes are formulated using the less expensive aromatic isocyanates. The yellowing that they undergo is quite noticeable. Therefore, one of the formulating rules for these plastics is "use only aliphatic isocyanates."

16.8.2   TABER ABRASION TESTING MEASURES MAR RESISTANCE

The light transmission (L/T) and haze of transparent plastics are key optical properties. Table 16.3 shows how the properties change after Taber abrasion.

TABLE 16.3   Abrasion Resistance Measured via Optical Properties

Test material	Initial light transmission and haze	After 50 cycles on Taber Abraser
Clad 4	L/T: 86.1–88.5%	L/T: 80.1–80.2%
	Haze: 3.8–4.5%	Haze: 30.9–36.0%
Clad 9	L/T: 88.0–91.0%	L/T: 75.4–83.4%
	Haze: 2.6–3.2%	Haze: 29.9–31.9%
Clad 8	L/T: 86.9–90.5%	L/T: 84.4–85.7%
	Haze: 7.6–6.8 %*	Haze: 20.0–22.7%

*A few fine bubbles increased haze reading.

Comments:

The Clad 4, 9, and 8 formulations are actually a series where the Teracol 1000 content increases from none in Clad 4 to 10% in Clad 9 to 20% in Clad 8. The haze readings increase due to Taber abrasion improved by the presence of the Teracol 1000. Clad 8 is the best of the series.

Formulating principle learned: Adding Teracol 1000 to a TP440/W system improves mar resistance of the transparent plastic.

16.8.3   SALT BLAST ABRASION RESISTANCE

Bare polycarbonate is easily made opaque due to scratches and abrasions. The cladding is expected to have superior abrasion resistance and thus extend the polycarbonate canopy's useful life. In order to simulate the erosive effects of rain, ice, or dust, transparent samples were blasted with salt in an empirical test and the light transmission and haze quantitatively measured. During our formulation development efforts, we tested dozens of samples using this technique. Space prohibits presenting all those data. Instead, a typical experiment involves this type of testing.

The Clad 9 formulation was made to be less cross-linked in a series of steps in order to make it more formable in the F-15 canopy-forming operation. We were interested in how that might affect abrasion resistance. Thus, samples of Clad 9 variations A, B, C, and D were blasted with salt particles and the haze and light transmission measured per ASTM 1044-73. These optical tests were measured after 0, 5, 10, and 20 blasts with salt at a 30° angle. The results are shown in Table 16.4.

TABLE 16.4   Light Transmission and Haze after Salt Abrasion of Clad 9-A through Clad 9-D

# of blasts	Property	Clad 9-A	Clad 9-B	Clad 9-C	Clad 9-D
0	Light transmission, %	91.0	90.8	91.3	91.0
0	Haze, %	0.5	1.1	1.0	0.8
5	Light transmission, %	90.6	90.2	89.4	86.5
5	Haze, %	4.7	5.8	7.9	14.0
10	Light transmission, %	88.7	87.2	87.1	83.6
0	Haze, %	9.8	15.0	19.0	34.4
20	Light transmission, %	87.9	85.4	80.9	78.4
20	Haze, %	11.3	21.1	34.6	49.3

Comments:

There was a marked inverse correlation between the degree of cross-linking and optical damage done. The order of most to least damaged is D > C > B > A.

These rigid polyurethanes had abrasion resistance intermediate between Plex 55 acrylic and polycarbonate. Plex 55 was more abrasion resistant than Clad 9-A and polycarbonate was slightly worse than Clad 9-D.

16.9 PROTECTION FROM NORMAL CHEMICALS

16.9.1 POLYMERIC CONSIDERATIONS

Polycarbonate is a remarkably tough, impact-resistant material, but it is vulnerable to many common chemicals such as cleaning agents. It will instantly craze from exposure to ketones, toluene, and other aggressive solvents and totally lose its transparency. The hard coat, which Sierracin applied to its F-15 canopies, served to protect the polycarbonate from hostile chemicals.

Our transparent, rigid polyurethane plastics are naturally solvent resistant because they are cross-linked polymers. Polycarbonate, on the other hand, is a linear polymer. Cross-linked polymers cannot be dissolved in solvents; however, they can swell if submerged in certain solvents.

16.9.2 SOLVENT RESISTANCE VERSUS DEGREE OF CROSS-LINKING IN THE CLADDING

Isopropyl alcohol after 15 hours caused no change in any of the series (Clad 9-A through D).

Methyl ethyl ketone (MEK) after 15 h submersion caused swelling and softening of the samples. The order of best to worst solvent resistance is Clad 9-A > B > C > D. Specimen D fared the worst actually breaking apart. Note that Clad 9-A is the most cross-linked of the series and Clad 9-D was the least cross-linked.

Polycarbonate would instantly craze on contact with MEK. Any of the clads would have protected the polycarbonate surface for the duration of a normal exposure.

16.9.3 CRAZE TESTING OF CLAD 8

Sierraclad™ samples of Clad 8-polycarbonate-Clad 8 were tested for crazing using Mil-P-25690A. A cantilever beam of the sample was stressed to 4000 psi and various test solvents applied to induce crazing. The test solvents were Isopropanol, methyl ethyl ketone, dichloroethane, cellosolve, and toluene. None of the solvents crazed the Clad 8 material.

16.10 TEMPERATURE SERVICE RANGE

16.10.1 DUROMETER VERSUS TEST TEMPERATURE

If you do not have expensive lab equipment, like a thermomechanical analyzer, a simple durometer and an oven and freezer to condition specimens will take you a long way. In order to establish some basic information on rigid polyurethane formulations, we examined the Shore D and A durometers of blended masterbatches. One was Clad 9-A, which was 178.24 parts TP-440 triol, 30.08 parts Teracol 1000, and 175.68 parts Hylene W diisocyanate. The second masterbatch was Clad 9-E, which was 60.16 parts Teracol 1000, 100.36 parts di-propylene gylcol, and 223.48 parts Hylene W.

These two masterbatches were blended in various proportions, catalyzed with DBTDL, and hardness specimens cast. The Shore durometers were measured at various temperatures. The D scale is appropriate for plastics and firm rubbers, whereas the A scale is appropriate for the entire range of rubber. Values near 0 or 100 end of the scales are meaningless, so both scales were used. The results are shown in the Table 16.5.

Table 16.5 Shore D Durometer of Various Transparent Materials versus Test Temperature

Material	75°F	100°F	150°F	210°F	300°F
Polycarbonate	74	74	73	71	65
Plex 55 Acrylic	82	82	80	75	45
Plex II Acrylic	80	80	78	75	38
Clad 5	74	74	73	71	Note 1
Clad 4	73	73	70	35	No data
Clad 8	70	66	40	35	34

Note 1: Highest reading was 43 D at 255°F.

16.10.2 TEMPERATURE RANGE OF THE CLAD 9-MODIFIED SERIES

The modified Clad 9-A–E series was similarly evaluated. The results follow:

TABLE 16.6   Shore D Durometer versus Test Temperature versus Formulation

Test temperature, °F	Clad 9-A	Clad 9-B	Clad 9- C	Clad 9-D	Clad 9-E
7	75	75	75	76	76
75	73	73	74	75	73
105	72	72	73	74	72
150	67	66	65	68	68
175	45	46	50	50	45
240	33	30	25	18	Melted
280	Split	28	25	15	Melted

Clad 9-A was the most cross-linked and split at 280°F. Clad 9-E was uncross-linked and it melted above 240°F. Clad 9-B, -C, and -D behaved similarly up to 175°F, but softened progressively with decreasing cross-linking at higher temperatures.

16.10.3 THERMAL BEHAVIOR OF CLAD 4 DURING CURING AND POST-CURING

The Clad 4 formulation was evaluated at 0.90, 0.95, 1.00, 1.05, and 1.10% of stochiometric (i.e., one isocyanate per one hydroxyl group). That is, we had been using 96.84 g of Hylene W for every 103.16 g of TP 440. Now, we varied the amount of Hylene W while holding the TP 440 amount constant. The term "isocyanate index" is also used to indicate how the NCO/OH ratio is varied.

We had made three sets of durometer specimens. We cured them all at 1 h at 150°F and removed one set. Next, we post-cured the remaining specimens 1 h at 200°F. Then we removed another set. The final set was post-cured for 1 h at 250°F + 1 h at 330°F. The durometers were measured and observations made.

The durometer specimens all remained clear and transparent during the entire cure cycle. However, the specimens turned from very slightly yellow to yellow after the 330°F exposure. The results are shown in the Table 16.7.

TABLE 16.7 Shore D Durometer versus Isocyanate Index versus Step Cure Temperature

Cure schedule	0.90	0.95	1.00	1.05	1.10
1 h at 150°F	73 D	73 D	73 D	73 D	73 D
Plus 1 h at 200°F	73 D	75 D	75 D	75 D	74 D
Plus 1 h at 250°F	75 D	74 D	75 D	74 D	75 D
Plus 1 h at 330°F					
Final color	Yellow-orange	Yellow-orange	Yellow	Yellow	Yellow

Comments:

1. Clad 4 is a highly cross-linked plastic, and as such shows a constant Shore D reading over differing NCO/OH ratios.
2. Additional cross-linking occurs at the higher postcuring temperatures. Available free isocyanate can undergo allophanate and biuret reactions at higher temperatures. Inability to further cross-link due to lack of free isocyanate may explain why the 0.90 and 0.95 index specimens turned yellowish orange. They were not as highly cross-linked as the others.

16.11   BIRD STRIKE PROTECTION

16.11.1   BIRD IMPACT AND THE SIERRACIN FALLING DART TEST

Actual windshields are tested for their impact resistance by firing a pre-weighed chicken from a special cannon into the windshield and assessing the damage. Obviously, this empirical test provides assurance that a vendor's windshields and canopies were safe to fly. However, the bird impact test was far too expensive to use to screen materials. Sierracin had devised their own empirical test to use as a fast comparison of different transparency designs. One-quarter inch thick specimens cut to 6 × 6 inch squares were placed on a steel support with a 4 inch diameter circle cut out of it. A steel ram having a 1 inch diameter rod with a hemispherical end was used to impact the supported specimen. A 50 pound weight was attached to the ram and the weighted ram raised various heights above the specimen and dropped. The condition of the specimen determined after the drop. If the specimen survived the test drop, a higher drop height was tried. This iterative process was repeated until the specimen failed or the maximum height of 6.5 feet failed to damage the specimen.

16.11.2   COMPARATIVE FALLING DART IMPACT RESULTS

The tremendous difference between the impact resistance of acrylic plastics and that of polycarbonate is illustrated in the following table. Acrylic plastics shatter at about 20 ft-lb of energy, whereas polycarbonate plastics are still structurally sound after being struck with 300 ft-lb of energy. Fusion-bonded acrylic/polycarbonate composites shattered at 75 ft-lbs. Certain Sierraclad™ composites were impact resistant as monolithic polycarbonate. The name "Sierraclad" was given to the concept of rigid polyurethane facings adhering to polycarbonate.

TABLE 16.8   Falling Dart Impact Results

Material Description	Drop height, energy ft-lb	Comments
Polycarbonate, monolithic,		
At 75°F:	6.5 ft, 330	Makes a dent
At −65°F	6.5 ft, 330	Makes a dent
Plex 55 Acrylic, monolithic	0.4 ft, 20	Shatters
Monolithic Clad 5	0.4 ft, 21	Shatters
Monolithic Clad 4	13 in, 30	No damage at 13 inches Shatters at 15 inches
Monolithic Clad 9	2 ft, 100	Dimple
Monolithic Clad 8	6 ft, 300	Dimple
Sierraclad: Clad-4/PC, clad up		
At 75°F:	6.5 ft, 330	Dimple
At −65°F:	6.0 ft, 330	Dimple, fails 6.5 ft
Sierraclad: Clad-8/PC, clad up		
At 75°F:	6.5 ft, 330	Dimple
At −65°F:	6.5 ft, 330	Dimple
Sierraclad: Clad-9/PC, clad up		
At 75°F:	6.5 ft, 330	Dimple
At −65°F:	6.5 ft, 330	Dimple
Sierraclad: Clad 5, clad up	3.0 ft, 150	Dimple

Unless otherwise stated, the clad side was up during the drop test. The readings are generally worse with the clad side down.

16.11.3 FORMULATING TO IMPROVE DART IMPACT RESULTS

16.11.3.1 VARIED DIOL MODIFIERS

An experiment was conducted to see the effect that long chain diols had on improving dart impact results. Table 16.9 shows the trial formulations. Each had a different diol, but all were assessed at 20% of the polyol blend level.

TABLE 16.9 Composition of Diol-Containing Rigid Polyurethane Trials

Trial name	Formulation	Diol name	Diol description
1	Pluracol TP-440 80 pbw Teracol-1000 20 pbw Hylene W 82 pbw	Teracol 1000	1000 mol. wt. poly-(oxybutylene) diol
2	Pluracol TP-440 80 pbw Teracol-2000 20 pbw Hylene W 79 pbw	Teracol 2000	2000 mol. wt. poly-(oxybutylene) diol
3	Pluracol TP-440 80 pbw PPG-1000 20 pbw Hylene W 82 pbw	PPG 1000	1000 mol. wt. poly-(oxypropylene) diol
4	Pluracol TP-440 80 pbw PPG-2000 20 pbw Hylene W 79 pbw	PPG 2000	2000 mol. wt. poly-(oxypropylene) diol

16.11.3.2 TEST RESULTS

The trials were catalyzed, cast, and cured as 1/4 inch monolithic sheets. The falling dart impact results are shown in Table 16.10.

TABLE 16.10   Falling Dart Impact Results

Trial	Drop height survived, ft	Equivalent energy, ft-lb	Comment
1	3.5	175	Shattered at higher drop heights
2	6.5	325	One specimen shattered
3	4.0	200	One specimen shattered
4	4.5	225	Shattered at higher drop heights

16.11.3.3   DART IMPACT CONCLUSIONS

1. Trial 2, containing Teracol-2000, was the best impact-resistant formulation. It was about twice as resistant as seen with trial 1 containing Teracol 1000.
2. A similar increase in impact resistance going from 1000 to 2000 molecular weight did not occur with the polyoxypropylene diols. Perhaps, the difference is that polyoxybutylene chains are stereoregular, crystallizable chain segments, whereas polyoxypropylene chain segments are not stereoregular.

16.11.4   THE THERMAL STABILITY OF DIOLS

In another experiment, where a PEP 550/W masterbatch was blended with a P300 diol/W masterbatch, the P300 diol was responsible for causing yellowing of the resultant transparent plastics. Dow P-300 was a 300 molecular weight diol based on a polyoxypropylene chain. The series of intermediate blends were highly exothermic. In fact, they were so exothermic that the yellowing was from thermal degradation of the diol. This observation reminded me of the poor heat stability of polyoxypropylene-based elastomers. The Teracol diols were preferable because they have the more stable polyoxybutylene chains, are stereo-regular, and have a lower glass transition temperature.

16.11.5   ACTUAL BIRD STRIKE TESTING OF SIERRACLAD™

Three double clad A/T 37 windshields were manufactured and bird tested at 300 mph. The cladding was Clad 8. All three windshields survived the test.

16.12 CANOPY FORMABILITY

16.12.1 THE CONCEPT OF POLYMER LINEARITY

Sierracin had developed great expertise in forming the windshields and canopies of numerous aircraft. They had facilities for producing their own stretched acrylic plastic, and they used Lexan polycarbonate for canopies needing high impact resistance. Both of these transparent plastics are composed essentially of linear polymers. It is true that stretched acrylic has a small degree of cross-linking, but it also has many linear chains. The original rigid polyurethanes, which I first showed to Dave Voss, were highly cross-linked thermosets. So, there is a fundamental difference here between my polyurethanes and the usual canopy-forming plastics. Linear polymer chains depend on interchain attractive forces to maintain their current shape. As temperatures are raised to the softening temperature, those forces diminish and the linear chains can slide past each other as the main mechanism for the plastic mass to assume a new shape. Upon cooling to ambient temperatures, the locked-in stresses in the formed canopy are minimal.

However, in the case of our polyurethane three-dimensional polymer networks, there are numerous cross-links which tend to pull the stretched chain segments back to the original shape of the mass. At the forming temperature for polycarbonate, the polyurethane cladding has transformed into a rubbery mass. It can be deformed so much and not anymore without breaking. Upon cooling, the new configuration is locked in although the stretched chain segments are still highly stressed. Loss of adhesion to the polycarbonate is also a potential problem from the stresses.

Reformulation of the rigid polyurethanes to attain more of a linear polymer nature and less of a three-dimensional network nature is the primary goal of the following experiments. However, we need to maintain the same rigidity as acrylic or polycarbonate. Rigidity of a polyurethane plastic is directly related to the percentage of isocyanate in the formulation. In order to accomplish both rigidity and increased linearity, we needed to look into short-chain diols to react with our current diisocyanate (Hylene W). The more formable formulations will be significantly different from the original polyurethanes and may exhibit different properties than them.

16.12.2   GENERAL PROCEDURE FOR DOME-FORMING EXPERIMENTAL SERIES

A series of formulation trials were undertaken to develop a Sierraclad panel capable of undergoing forming into a canopy configuration without the cladding breaking or delaminating from the polycarbonate. The test panels were made 24 inch × 24 inch in area. The procedure was as follows:

1. 0.25 inch thick polycarbonate sheets were cut to 24" × 24", primed with 647-17, and dried 20 h at 250°F.
2. Two foot square glass buffers were lined with polypropylene film and heat to produce a tight, smooth surface.
3. Tygon tubing was applied to the periphery of the buffers to form a leak tight seal.
4. Experimental Clad mixture was cast between polycarbonate and glass buffer. Clamps were tightened against 30 mil spacers to produce a 30 mil thick cladding.
5. Stepwise heat cure schedules were employed to cure the clad material.
6. The glass buffer(s) were removed and the panel post-cured if warranted.
7. The following specimens were taken from the panel:

TABLE 16.11   Test **Specimens Layout from the Sierraclad™ Panel**

Type	Quantity	Test specimen description	Purpose
A	1	Octagon (13" across)	Form a 9" diameter dome evaluation
B	1	4" × 4" square	Taber abrasion testing
C	3	5.5" ×5.5" square	Falling dart impact testing
D	2	6.5" × 6.5" square	Form a 9" diameter dome evaluation
E	5-10	Irregular cut off pieces	Salt blast, adhesion, weathering

16.12.3   INITIAL STATE OF THE ART IN POLYURETHANE PLASTIC DOME FORMING

Clad 4 was a highly cross-linked polymer utilizing a 400 molecular weight triol TP-440 as the sole polyol. It was unsuitable for deep dome forming due to its limited ability to elongate. Clad 9 is a modification of Clad 4, where 10% of the TP-440 is replaced with 100 molecular weight diol,

Teracol 1000. It was also unsuitable for dome forming. What was needed was to modify the Clad 4 or Clad 9 by making them less cross-linked and more linear, so that polymer chains can slide past each other.

In order to gain greater extensibility, a host of short-chain diols were considered. Dipropylene glycol and tri-ethylene glycol were evaluated in the following experiments.

16.12.4 LINEAR POLYURETHANE POLYMER INVESTIGATION

Prior to incorporating short-chain diols into Clad formulations, we wanted to see how blending masterbatches of 32.68 pbw di-propylene glycol + 67.32 pbw Hylene W (called DPG-W) and 78.37 pbw Teracol 1000 + 21.63 pbw Hylene W (called T-1000-W) behaved. These polymers should be almost 100% linear polymers. The following table shows the Shore D durometer of the different blends at five different temperatures:

TABLE 16.12   Shore D Durometer versus Temperature of Linear Polymeric Blends

Formulation	10°F	75°F	150°F	200°F	250°F
100% DPG-W	76	75	71	68	40
90% DPG-W + 10% T-1000-W	76	75	70	62	25
80% DPG-W + 20% T-1000-W	76	72	63	40	16
50% DPG-W+ 50% T-1000-W	70	35	23	15	12
100% T-1000-W	13	11	9	0	0

Comments:

(1) The DPG-W plastic had the best thermal range, whereas the T-1000-W material was a rubber across the thermal range. The structure for DPG is shown below:

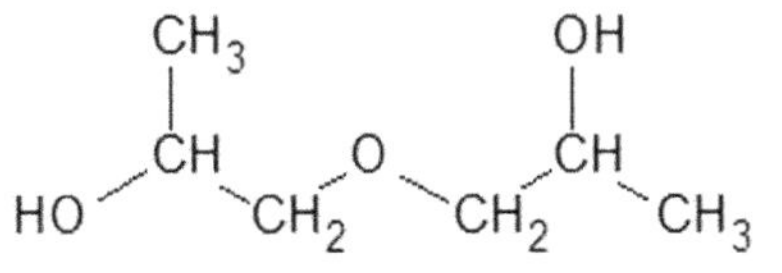

FIGURE 16.2   Di-propylene glycol.

(2) The 50% DPG-W+ 50% T-1000-W material was a rubber at room temperature and higher.

16.12.5   DECREASED CROSS-LINKING VIA DI-PROPYLENE GLYCOL

16.12.5.1   CLAD 9 WITH INCREASED LINEARITY

Formulations were prepared which decreased the triol TP440 in steps, while increasing the Teracol 1000 and di-propylene glycol in steps. The short-chain diol helps maintain equivalent durometer. The following table shows the formulations:

TABLE 16.13   Formulation Modifications of Clad 9 to Decrease Cross-linking

Ingredient	Clad 9-A	Clad 9-B	Clad 9-C	Clad 9-D	Clad 9-E
Pluracol TP-440 (triol)	178.24	133.68	89.12	44.56	0.00
Teracol 1000 (diol)	30.08	37.58	45.08	52.64	60.16
Di-propylene glycol (diol)	0.00	25.10	50.20	75.27	100.36
Hylene W (diisocyanate)	175.68	187.64	199.60	211.53	223.48
Total weight	384.00	384.00	384.00	384.00	384.00

Note 1: Formulas Clad 9-A and Clad 9-E were masterbatches. Clad 9-B= 75% Clad 9-A +25% Clad 9-E, Clad 9-C = 50% Clad 9-A + 50% Clad 9-E, Clad 9-D = −25% Clad 9-A + 75% Clad 9-E.

Note 2: Two drops of DBTDL per added to each 100 gof mixture to catalyze.

Note 3: Cure was 2 h at 150°F plus 2 h at 200°F.

16.12.5.2   TENSILE TEST RESULTS OF CLAD 9-MODIFIED SERIES

Dome forming was not attempted on this series, but the tensile properties were tested to verify that elongation (stretch) had increased as the linearity of the polymer increases. The tensile properties of the same Clad 9 series are shown below:

TABLE 16.14   Tensile Properties of Polyurethane Formulations at 75°F and 300°F.

Formulation	Tensile strength, psi at 75°F	Tensile modulus, psi at 75°F	Ultimate elongation, %, at 75°F	Tensile strength, psi at 300°F	Ultimate elongation, %, at 300°F
Clad 9-A	6120	260,000	6.3	170	11
Clad 9-B	6900	290,000	29.0	160	12
Clad 9-C	7720	300,000	41.0	180	21
Clad 9-D	7980	300,000	137.0	370	82

Comments:

It is interesting that both tensile strength and ultimate elongation increase as the level of cross-linking is decreased (going from Clad 9-A to Clad 9-E). The casting for Clad 9-E was unsuitable for tensile testing due to cracks and bubbles which developed during the 200°F postcure.

16.12.6 DECREASED CROSS-LINKING VIA TRI-ETHYLENE GLYCOL

The following experiment had been conducted with less diol content to reduce cross-linking. Good dome-forming results were obtained at the most linear of three formulations. The following experiment extends that approach.

16.12.6.1 MORE LINEAR FORMULATIONS

Three formulations were prepared which decreased the triol TP440 in steps, while increasing the Teracol 1000 and tri-ethylene glycol in steps. The short-chain diol helps maintain equivalent durometer. The following table shows the formulations:

TABLE 16.15 Formulations with Decreased Cross-Linking at Constant Durometer

Ingredients	Clad 301	Clad 302	Clad 303
Pluracol TP440	14.79	12.42	10.05
Tri-ethylene glycol	17.98	19.55	21.11
Teracol 1000	13.69	13.88	14.08
Eastman RMB	2.00	2.00	2.00
Irganox 1010	2.00	2.00	2.00
Hylene W	49.54	50.15	50.76
Totals	100.00	100.00	100.00

Note: The polyol mix of trial 3 was hazy. However, after the addition of Hylene W, the resultant polymer was clear and transparent. The structure for tri-ethylene glycol is shown below:

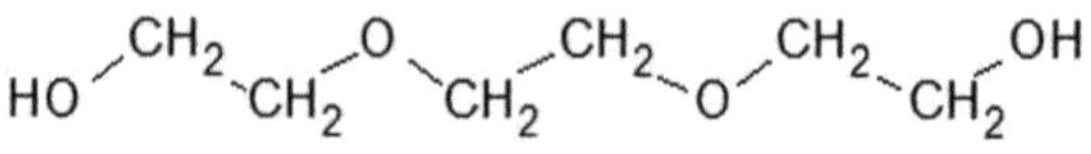

FIGURE 16.3 Tri-ethylene glycol.

16.12.6.2 DOME-FORMING EXPERIMENTAL RESULTS

The following table shows the results of forming 4 inch diameter and 9 inch diameter domes where the Sierraclad cladding was one of the formulations above:

TABLE 16.16 Results of Forming Sierraclad™ 4" and 9" Domes

Clad name	Dome diameter, inch	Depth, inch	Stretch, %	Comments
301	4	1.5	13.6	Failed in cool down, failed in adhesion
302	4	1.2	15.0	Clad cracked at 1.1 inch depth
302	4	1.3	16.3	Good dome, passed pull-back test
303	4	1.5	25.5	Good dome, passed pull-back test at 260°F
301	9	3.2	20.0	Okay until 3.2 inch depth, then broke
302	9	2.5	12.2	Cracked in forming
303	9	3.9	27.9	Good dome at 3 inch depth, broke at 3.9 inch

Comments:

1. Stretch is $(L_1 - L_0) \times 100\%/L_0$, where L_1 is the arc length, and L_0 is the original length.
2. The dome-forming results are not highly consistent, but the three formulations seem to be similar to one another. Shallow domes are doable, whereas deeply formed domes are not.
3. In other testing of Clads 301, 302, and 303, we learned they are essentially the same hardness, 70-72 D as we intended. The falling dart impact test results (around 6 ft drops) indicate that the increased linearity progressively helped impact results.

16.12.7   DOME-FORMING CLAD 320-MODIFIED FOR BETTER IMPACT

16.12.7.1   MASTERBATCHES:

In this experiment, the goal was to modify the dome-forming Clad 320 to improve its impact properties. Clad 320 became one masterbatch and Clad 330 was formulated to be the modifying masterbatch. Their formulations follow:

TABLE 16.17   Clads 320 and 330 Formulations

Ingredients	Clad 320	Clad 330
TP-440	16.71%	
Di-propylene glycol	15.69%	
Teracol 1000	12.22%	75.24%
RMB	2.00%	2.00%
Irganox 1010	2.00%	2.00%
Hylene W	51.38%	20.76%

16.12.7.2   BLENDS

Blended formulations per the following table were prepared. A total of 2 drops of DBTDL were used per 100 g of mixture. Trials were cast 30 mils thick on one side only of 0.30 thick polycarbonate and cured 2 h at 150°F plus 2 h at 200°F. The glass molds were removed and the parts post-cured for 8 h at 280°F.

TABLE 16.18   Formulations of Clad 320, Clad 321, and Clad 322

Ingredient	Clad 320	Clad 321	Clad 322
Clad 320 less Hylene W	48.62 pbw	46.18 pbw	43.76 pbw
Clad 330 less Hylene W		3.96 pbw	7.92 pbw
Hylene W	51.38 pbw	49.86 pbw	48.32 pbw

16.12.7.3   DOMES

The dome-forming results are presented in the Table 16.19.

TABLE 16.19   Results of Forming Sierraclad 4" and 9" domes

Clad name	Dome diameter	Appearance	Stretch, %	Pull-Back Test
320	4 inch	Excellent	16.3*	Passed
321	4 inch	Excellent	23.6	Passed
321	4 inch	Delaminated	37.0	Not run
322	4 inch	Excellent	19.5*	Passed
320	9 inch	Excellent	20.4**	Not run
321	9 inch	Excellent	17.6	Not run
322	9 inch	Excellent	16.0***	Not run

*Average of two domes.

**20.4% stretch correlates to a 3.75 inch deep dome.

***Okay at 3 inch deep (16% stretch), failed at 3.5 inch depth.

16.12.7.3   OTHER TESTING:

The goal of improving the impact resistance was moderately successful. The clad up orientation passes 6.5 ft on the dart impact test as expected. The clad down orientation is the harder orientation to pass. Clad 320 passed 6 inches, Clad 321 passed 1 foot, and Clad 322 passed at 2 feet drop height.

16.13   WEATHERABILITY

16.13.1   PROBLEMS WITH THE A-37 SIERRACLAD™ WINDSHIELDS

We were alerted to a problem with one of the Sierraclad A-37 windshields in service in July 1976. It was a double clad construction using Clad 8 cladding. The exterior clad surface had deteriorated to where cleaning it with isopropanol caused a smear. The windshield had only been in service for 4 months, but had been routinely parked facing the sun. Another older windshield of the same construction was still serviceable. The cladding on the inside surface was still pristine. We had an outdoor weathering problem with the Sierraclad™ concept.

16.13.2 SCREENING OF UV ABSORBERS AND ANTIOXIDANTS

An extensive study of UV absorbers and antioxidants was undertaken initially focused on Clad 8, which is the cladding on the UV-damaged A-37 windshield. These additives were examined as sole additives or in combination, and at different levels. The following tests were performed:

Test # 1: Observe color and transparency initially, and after 6 h at 300°F, and 30 h at 300°F.

Test # 2: Conduct weatherometer aging per FTMS-406 method 6024. Periodically, remove specimens and measure light transmission and haze after Taber abrasion.

The best additive combination was 2% Eastman RMB UV absorber plus 2% Irganox 1010 antioxidant. Eastman RMB is identified as resorcinol monobenzoate.

The addition of these additives typically extended the life of the cladding five-fold.

16.13.3 OUTDOOR WEATHERING

Samples were prepared for outdoor weathering using Clad 4, 5, and 9 with best UV absorber and antioxidant, based on weatherometer results. Samples were outdoor aged at three locations:

(1) Sylmar, California for convenience of observation
(2) Miami, Florida for heat and humidity type of climate
(3) Phoenix, Arizona for dry heat and solar radiation effects

16.13.4 WEATHERING PROJECT ENDED

The entire Sierraclad™ project was terminated before the weathering study was completed. A simpler approach to the F-15 canopy problem was undertaken instead.

16.14   SUMMARY

The original Clad 4, 5, 8, and 9 were highly cross-linked polymers. Their resistance to solvents, abrasion, and weathering were quite good. The impact resistance of the Clad 8, which had reduced cross-linking, was comparable to polycarbonate. Clad 4, 5, and 9 ranged from slightly better to much better than Plex 55 Acrylic plastic. However, these early claddings did not have the high tensile elongation needed for physically transforming from a flat sheet to the shape of a fighter jet canopy. Various formulation modifications were made to optimize the Clads for deep draw forming. The more linear that we made the polymer, the better its tensile elongation properties and the better its ability to form a dome. However, making the polymer more linear reduced resistance to solvents, abrasion, and weathering. We added UV absorbers and antioxidants to enhance weathering resistance, but it did not help enough.

Eventually, it became evident that polyurethane plastic cladding was not going to fulfill all of the requirements of the F-16 canopy application. On one end of the formulating spectrum, we could attain outstanding weather ability and abrasion resistance. However, these properties were obtained by sacrificing impact resistance and formability, whereas at the other end of the spectrum, we could match the outstanding impact resistance of polycarbonate and have a material capable of deep draw thermoforming. However, these properties were obtained at the expense of poor weather ability and abrasion resistance. It appeared increasingly unlikely that we would find a way to have all the desired properties in a single formulation. Even if we had a more ideal formulation, there was still the problem of mold releasing glass buffers that we had set aside. The Sierraclad approach was abandoned for a more immediate solution.

16.15   POSTSCRIPT

Dave Voss had learned that one of our competitors was producing acceptable F-16 canopies by marrying acrylic facings to a polycarbonate core ply via a flexible polyurethane interlayer. I was directed to see what I could come develop in the way of a polyurethane interlayer. Several weeks later, we had one. It was produced in a thick film form. A flat stack of acrylic facing, my interlayer, polycarbonate, my interlayer, and acrylic facing was assembled and placed in a vacuum bag. A large platen press applied heat and pressure to the assembly. A strongly bonded transparent composite was produced. The laminate was formed into F-15 canopies successfully. Large orders started coming in for this design. My interlayer was treated as a trade secret rather than going the patent option. Consequently, I will give no details on how we developed it.

16.17 PATENT

U.S. Patent 4,045,269 titled, "Transparent, formable, polyurethane, polycarbonate lamination" was awarded to David L. Voss, William A. Miller, and Ralph D. Hermansen on August 30, 1977. The application had been filed on December 22, 1975.

16.18 PRESENTATIONS

During the Conference on Aerospace Transparent Materials and Enclosures in Atlanta during November 18–21, 1975, Dave Voss presented a paper entitled "Polycarbonate Protection." It was a well-researched presentation of the history of different approaches to providing polycarbonate protection. He pointed out that acrylic facings laminated to polycarbonate with a rubbery interlayer had been working well for subsonic aircraft, but appeared to be not suitable for supersonic aircraft such as the F-15. He then introduced the "Sierraclad™" concept and discussed the progress we had made to date. He explained that the rigid polyurethane cladding could be varied over a wide range of properties by Sierracin's chemists.

KEYWORDS

- **Sierraclad**
- **cladding**
- **dome forming**

PART VII
Three Niche Technologies

CHAPTER 17

FLEXIPOXY TECHNOLOGY

CONTENTS

In this section of the book, we shall discuss three niche polymer technologies, which evolved during the compound development projects described in the previous pages. These niche technologies have importance because of their potential to help solve future material development projects. The three niche technologies are: (1) Flexipoxy, (2) Hacthane, and (3) Transparent polyurethanes. We do not claim to be the first to explore these niche technologies. Far from it, for example, much of the Hacthane technology already existed for electrical applications. Our contribution was to extend its usefulness. Flexipoxy technology had been delved into by others, but I think we made a lot of pioneering advances and consolidated the technology. The transparent polyurethanes technology evolution was totally unplanned, but an assignment caused me to use all my polymer knowledge to formulate a novel material. A spectrum of potential products resulted.

17.1   WHAT IS FLEXIPOXY?

Flexipoxy is a name I invented to help sell a development concept, which I knew was likely to be misinterpreted. I had an assignment to find or develop a thermal transfer adhesive in film adhesive form, which would be stored as a frozen premix. The adhesive had to be flexible and reworkable so I needed a weak elastomer. Polysulfides had been ruled out due to tendency to outgas and polyurethanes would not work for the film adhesive process due to foaming. I thought epoxies might work in the application, if I could formulate a rubbery one. Experience had taught me that most people associate the word "epoxy" with a hard, brittle plastic. Consequently, I needed a different word than "epoxy" to describe our rubbery epoxy compounds to avoid that automatic association with hard and brittle materials. The word "Flexipoxy" did the trick beautifully.

The following table examines the considerations involved in that polymer selection.

Flexipoxies are formulated by using flexible epoxy resins at the 100% level with flexible curatives. No epoxy resin manufacturer advised using flexible epoxy resins as lone resins, but did advise using them in conjunction with rigid epoxy resins in order to make the resultant plastic less brittle. We took a bold step and used them as sole resins in our formulations. After extensive screening, we found useful formulations for our thermal transfer adhesive application.

Our Flexipoxy materials are rubbers or elastomers because their glass transition temperatures are well below room temperature. By contrast, plastics have their glass transition temperatures well above room temperature. Like other elastomers, Flexipoxies have the ability to be stretched and snap back to their original length when released. They have a Poisson's ratio very near 0.5, which means they maintain a constant volume when deformed. Like rigid epoxies, Flexipoxies are good adhesives to a wide variety of substrates. They can be stored as one-component, frozen premix compounds for 6 months or more.

TABLE 17.1   In Situ Curing, Elastomeric Polymers

Elastomeric polymer	Advantages	Disadvantages
Flexipoxy	Versatility, good adhesion, stable frozen premixes, low outgassing	Limited mechanical strength, dermatitis
Polysulfides	Good adhesion, good solvent resistance, stable frozen premixes	Bad outgasser, rotten egg odor, sensitive to cure cycles. Bad outgassing
Silicones	Widest useful temperature range, room temperature cures	Contaminant (prevents adhesion), outgassing
Polyurethanes	Versatility, good adhesion, instable frozen premixes, wide selection of mechanical and electrical properties	Isocyanate allergy, moisture bubbling, few frozen premix stable

17.2   A BRIEF HISTORY OF THE FLEXIPOXY TECHNOLOGY EVOLUTION

The following is a brief recap of the Flexipoxy evolution of new compounds discussed in the book: Flexipoxy 100 thermal transfer film adhesive was the first Flexipoxy development and it solved an important problem (i.e., outgassing), which would affect getting future satellite orders. New formulating discoveries allowed us to improve the dielectric properties and thermal range over those of Flexipoxy 100 with Flexipoxy 208BF adhesive. However, the outgassing of Flexipoxy 208BF was very slightly high. So despite the other superior properties, it was not implemented. Yet, there is a wealth of formulating information about flexible epoxies disclosed in this project. Flexipoxy was also the right material to impregnate alumina particles in a Naval weapon application. Flexipoxy formulations were selected as potting compounds for evaluation as a Delco electronics project. They also served as a base to formulate flexible epoxy hot melts conformal coatings, which would cross-link after application to develop needed thermoset properties. As a participant in an NCMS consortium, we used our Flexipoxy formulating knowledge to develop a drop-resistant organic solder. None of the commercial suppliers were able to do it, but we did it. Finally, we learned how to make room temperature-stable flexible epoxy compounds, which would cure at reasonably low temperatures.

The following table lists the chapters, which discuss Flexipoxy custom-formulated compound development.

TABLE 17.2   Flexipoxy Development Projects and Where to Find Them

Flexipoxy topic	Book location
Development of Flexipoxy 100 thermal transfer film adhesive	Chapter 4
Expanded characterization of Flexipoxy 100 adhesive	Chapter 5
Development of a superior Flexipoxy thermal transfer film adhesive	Chapter 6
Development of a custom-formulated Flexipoxy impregnant	Chapter 8
Development of three kinds of RT-stable Flexipoxy adhesives	Chapter 11
Development of Flexipoxy encapsulants or potting compounds	Chapter 12
Development of Flexipoxy reactive hot melt conformal coatings	Chapter 13
Development of a Flexipoxy drop-resistant organic solder	Chapter 15

17.3   MAJOR BREAKTHROUGHS

There were five milestones or major breakthrough achieved and they are: (1) a versatile range of rubber-like hardness, (2) improvement in dielectric properties, (3) significant lowering of low temperature flexibility, (4) one-component, flexible epoxy RT-stable compounds, and (5) flexible epoxy reactive hot melts.

17.3.1   A VERSATILE RANGE OF RUBBERY HARDNESS

The following table shows the durometers of various epoxy resins/curative combinations. All of those with an "A" scale rating would be included in our concept of Flexipoxy. Notice that some Flexipoxies had Shore A readings of 25–35, which is very soft and flexible. Some had readings of 80–92 Shore A, which is a fairly firm rubber. All intermediate durometers are represented.

TABLE 17.3 Durometers of Several Epoxy Resin/Curative Combinations

Epoxy resin	With TETA	With Dytek A	With DP-3680	With ATBN
Epon 828	87D	85D	74D	69A
Epon 871	53A	37A	45A	35A
Epon 872	75D	75D	60D	67A
DER 732	60A	45A	56A	30A
DER 736	73A	69A	62A	35A
NC-514	77D	78D	65D	60A
NC-547	92A	81A	67A	25A
Heloxy 84	57A	20A	46A	47A

17.3.2 IMPROVEMENT IN DIELECTRIC PROPERTIES

Flexipoxy 100 adhesive was the best that we could do in the earliest days of Flexipoxy formulating. Its volume resistivity was only in the 10^{11} ohm-cm range at room temperature and fell two decades by 200°F. This was inadequate for some applications. Discovery of better flexible epoxy resins and curatives allowed us to improve volume resistivities by two or more decades. Dielectric constants and dissipation factors also improved significantly. Chemically, we found that the more hydrophobic the epoxy resin, the better the dielectric properties. The same principles apply to polyurethanes and presumably other polymers. The following tables show the volume resistivity of several Flexipoxy compounds.

TABLE 17.4 Volume Resistivity of Various Flexipoxy Compounds versus Test Temperature, in Ohm-cm

Test temperature °F	Flexipoxy 100 paste	Epon 872/ DP3680	NC-514/ DP-3680	NC-547/ DP-3680	Flexipoxy 175
75	1.6×10^{11}	1.9×10^{16}	5.3×10^{15}	1.2×10^{14}	3.6×10^{14}
120	3.6×10^{10}	1.3×10^{13}	2.8×10^{13}	1.3×10^{13}	2.0×10^{13}
160	8.5×10^{9}	6.4×10^{11}	1.8×10^{12}	1.8×10^{12}	1.8×10^{12}
200	4.8×10^{9}	1.3×10^{11}	3.5×10^{11}	2.7×10^{11}	2.7×10^{11}

17.3.3 SIGNIFICANT LOWERING OF LOW TEMPERATURE FLEXIBILITY

As the temperature of an elastomer is lowered, it remains rubbery until it transitions, first into a leathery material and finally into a hard rigid material. The midpoint of the leathery transition is called the glass transition temperature. The table below shows how much it can be lowered using the best choice of epoxy resin and curative.

TABLE 17.5   Glass Transition Temperature of Several Flexipoxy Compounds

Flexible epoxy resin	Flexible curative	Tg, °C
Cardolite NC-514	DP-3680	0.0
Heloxy 67	DP-3680	−15.3
Epon 871	DP-3680	−26.6
Epon 871	Dytek A	−28.3
Heloxy 84	ATBN	−34.3
DER 732	DP-3680	−35.4
Heloxy 84	Dytek A	−41.8
Heloxy 505	DP-3680	−43.0
Heloxy 84	DP-3680	−44.2
Heloxy 505	Jeffamine D2000	−44.3
Cardolite 547	DP-3680	−60.0

17.3.4 ONE-COMPONENT, FLEXIBLE EPOXY RT-STABLE COMPOUNDS

The details are told in Part II Chapter 9, but we made a big advance in the usefulness of Flexipoxies by demonstrating that they can be formulated into one-component, room temperature-stable compounds. Although the technology for making room temperature-stable rigid epoxies was well established, no one seemed to have done it for flexible epoxies. We showed that flexible epoxies can be formulated into thermal transfer adhesives, electrically conductive adhesives, and general-purpose adhesives.

Such compounds are ideal for automated dispensing and mass production. They have an advantage over frozen premix compounds in not requiring a thaw period.

17.3.5 FLEXIBLE EPOXY REACTIVE HOT MELTS

We also did development work to assist Delco Electronics Corporation regarding reactive hot melt compounds. Significant progress was made in developing reactive hot melt flexible epoxies for them. It would be easy to develop a rigid epoxy hot melt because many solid epoxy resins are commercially available. As for solid flexible epoxy resins, they are virtually nonexistent. The key to our success was formulating epoxide-terminated prepolymers using flexible epoxy resins. The details are in Part III Chapter 2.

17.4 APPLICABLE FLEXIPOXY PATENTS

TABLE 17.6 Patents Applicable to Flexipoxy Development

U.S. Patent	Date granted	Description
4.866,108	September 12, 1989	Flexible epoxy, thermal transfer film adhesive for flatpacks
5,367,006	November 22, 1994	Optimized flexible epoxy paste or film thermal transfer adhesive
5,457,165	October 10, 1995	Flexible epoxy encapsulants or potting compounds
5,510,138	April 23, 1994	Reactive flexible epoxy hot melt conformal coatings
5,575,956	November 19,1996	RT-stable, electrically conductive, flexible epoxy adhesives
5,929,141	July 27,1999	Drop-resistant, electrically conductive, flexible epoxy adhesives
5,965,673	April 23, 1994	Reactive hot melt, flexible epoxy-terminated adduct
6,060,539	May 9, 2000	RT-stable, thermally conductive, flexible epoxy adhesives

TABLE 17.6   *(Continued)*

U.S. Patent	Date granted	Description
6,132,850	October 17, 2000	Reworkable, thermally conductive, flexible epoxy impregnant
6,723,803	April 20, 2004	RT-stable, general-purpose, flexible epoxy adhesives

17.5   FLEXIPOXY PUBLICATIONS

(1) "The Development of Custom-Tailored, Thermal Transfer Adhesives" by Ralph D. Hermansen and Steven A. Tunick at the Third International SAMPE M & P Electronics Conference in June 1989.

(2) "Development of a Thermal Transfer Adhesive For Space Electronics" by Ralph D. Hermansen, Robert B. Mitsuhashi, James C. Cammarata, and Matthew T. Mika. I presented the paper in June 1990 at the Fourth International SAMPE Electronics M & P Conference in Albuquerque.

(3) "Space-Simulated Aging Methods for Electronic Assembly Adhesives" by Dr. Tom Sutherland and Ralph Hermansen. Dr. Tom Sutherland presented the paper at the Fifth International SAMPE Electronics M & P Conference in Los Angeles on June 20, 1991.

(4) "The Evolution of Space Qualified Thermal Transfer Adhesives" by Ralph D. Hermansen, Steven E. Lau, and Robert B. Mitsuhasihi. I presented the paper in June 1992 at the Annual International SAMPE Electronics M &P Conference in Baltimore.

(5) "Advances in Custom-Formulated Flexible Epoxies" by Ralph D. Hermansen and Steve Lau, was run in Adhesives Age Magazine, July 1993. This same paper was presented as a vugraph presentation by Ralph D. Hermansen at the Society of Plastic's Industry—Epoxy Resin Formulator's Spring Conference at Newport Beach, California in March 1993.

KEYWORDS

- **Flexipoxy**
- **hot melts**
- **flexible epoxy resins**

CHAPTER 18

HACTHANE TECHNOLOGY

CONTENTS

18.1   WHAT IS HACTHANE?

The word "Hacthane" was one I selected by combining the segment "Hac" for Hughes Aircraft Company with "thane" from urethane. "Hacthane" would henceforth be our designation for polyurethane formulations developed in out lab. Hacthane 100 was our first compound. It was developed to be a thermal transfer adhesive for space electronics. This adhesive had been sought for years unsuccessfully. The secret to our success was finding a way to formulate a polyurethane adhesive, which could be packaged as a one-component frozen premix and stored at freezer temperatures for 6 months or more.

18.2   A BRIEF HISTORY OF THE HACTHANE TECHNOLOGY EVOLUTION

Hacthane 100 thermal transfer adhesive was unique, that is, it was non-outgassing, made a stable frozen premix adhesive, was thermally conductive, had excellent dielectric properties, and worked as well or better than the out-gassing adhesive, it replaced. The key to the excellent electrical properties was its polyol, namely, PolyBd R45HT. This butadiene-type polyol also provided a low glass transition temperature. Some modifications to this adhesive were made to allow it to adapt to other applications within Hughes Aircraft Company.

For example, Hacthane 108 thermal transfer adhesive was a version of Hacthane 100, especially modified for very easy flatpack removal. It had one-third the lap shear strength of Hacthane 100.

For another example, a version designated Hacthane 120 adhesive was developed for the Nightsight program. The engineers at Radar Systems Group needed a thermally conductive adhesive to bond an aluminum heat sink to the focal plane array. The desired adhesive needed to be more conductive than Hacthane 100 and cure faster to accommodate a high-volume production rate. Hacthane 120 adhesive has a thermal conductivity of 0.52 BTU/Hr Ft °F, passes NASA outgassing requirements, is hydrolytically stable, and is very fast processing.

Hacthane 121 was developed for RSG as part of a Hacthane IR & D project. Hacthane 121, a slower curing version of Hacthane 120, was characterized in order to input its properties into a computer simulation program. The adhesive had favorable results.

Hacthane 303 was developed for the Hughes Radar Systems Group as a flow-under adhesive. This unfilled adhesive wicks under flat-bodied components when the assembly is put under vacuum. The thermal conduction of the cured polymer is better than that of air.

Hacthane 220 adhesive was developed as a candidate for the Comlite program for the Hughes Radar Systems Group. It was both electrically and thermally conducting.

The next large Hacthane development effort was with Delco Electronics Corporation, our sister organization under Hughes Electronics. Hacthane 110 potting compound represented the best compound for an air bag sensor potting application. An extensive formulation improvement project was conducted over a 2-year span. Hacthane 115 was the optimized result. It had superior adhesion and stabilization against loss of properties in the severe under-the-hood environment.

Another project funded by Delco Electronics Corporation, was the development of reactive hot melt conformal coating material. Although we invented Flexipoxy compounds to solve this problem, our polyurethane solutions were even better.

The following table lists the chapters, which discuss Flexipoxy custom-formulated compound development.

TABLE 18.1 Hacthane Development Projects and Where to Find Them

Hacthane topic	Book location
Development of Hacthane 100 thermal transfer filleting adhesive	Chapter 3
Expanded characterization of Hacthane 100 adhesive	Chapter 5
Development of Hacthane 115, an optimized encapsulant	Chapter 12
Development of polyurethane reactive hot melt conformal coating	Chapter 13

18.3 MAJOR BREAKTHROUGHS

There were four major breakthroughs from our Hacthane development experience. They are: (1) development of a polyurethane elastomeric formulation, which if made into a frozen premix was stable for 6 or more months of freezer storage, (2) an optimized encapsulant or pottant for severe environments, (3) an optimized encapsulant or pottant for adhesion, and (4) an optimized reactive polyurethane hot melt adhesive.

18.3.1   FROZEN PREMIX WITH MONTHS OF LIFE

Polyurethane elastomers, formulated with polybutadiene-based polyols, had high promise for solving the Hughes Space and Communications Group's quest for an ideal thermal transfer adhesive. However, the storage life of these adhesive in frozen premix form was only around 1 month. SCG needed 3 months minimum and preferred 6 months. I reasoned that aliphatic isocyanates, being less reactive than aromatic isocyanates, might extend the frozen premix storage life, and it worked better than I had imagined. The cost increase might have been prohibitive for commercial applications, but for aerospace applications, it was insignificant. The story does not end there. As we manufactured batch after batch of Hacthane 100 adhesive in our facilities, the storage life increased from 6 months to 9 months. Careful drying of the ingredients to eliminate water is an equally important factor in prolonging storage life.

18.3.2   OPTIMIZED ENCAPSULANT OR POTTANT FOR SEVERE ENVIRONMENTS

The biggest aging problem for the Hacthane 110 potting compound was undesirable changes due to dry heat aging. In particular, a hard skin developed on the air face of the potting compound. With longer exposure, the skin continuously thickened until the entire mass had become hard and brittle. An extensive evaluation of four different antioxidants at three different levels and at three different aging environments was undertaken to find the best stabilizer. Volume resistivity was found to be acceptable under all conditions. Durometer was a good indicator of the hardening phenomena. Tensile strength and ultimate elongation at the end of 4 weeks aging was very informative. Durometer over 90 Shore A correlate with a drastic drop in ultimate elongation, which means the potting compound is no longer adequately flexible.

The following table compares Hacthane 110 with Hacthane 110-M in the three aging environments. Hacthane 110-M has 1% addition of Irganox 1010 antioxidant added to Hacthane 110. Irganox 1010 was the best antioxidant from an extensive study. The optimized pottant, Hacthane 115 has 1% Irganox 1010 in it based on this and other data. The reader should notice the reduced hardening of Hacthane 110-M versus Hacthane 110 for each of the three environments. The stabilized Hacthane retains its flexibility much longer than the unstabilized Hacthane 110.

TABLE 18.2 Durometer and Ultimate Elongation versus Aging of Hacthane 110 Potting Compound

Test pottant, type aging	Initial	1 Week	2 Weeks	3 Weeks	4 Weeks
Hacthane 110	D = 56	D = 73	D = 76	D = 76	D = 76
Ambient aging	UE = 100%				UE = 82%
Hacthane 110-M	D = 54	D = 64	D = 63	D = 63	D = 63
Ambient aging	UE = 100%				UE = 69%
Hacthane 110	D = 56	D = 74	D = 75	D = 76	D = 77
85°C/95% RH	UE = 100%				UE = 89%
Hacthane 110-M	D = 54	D = 66	D = 66	D = 66	D = 66
85°C/95% RH	UE = 100%				UE = 90%
Hacthane 110	D = 58	D = 84	D = 92	D = 92	D = 92
125°C aging	UE = 100%				UE = 5%
Hacthane 110-M	D = 55	D = 69	D = 69	D = 72	D = 76
125°C aging	UE = 100%				UE = 22%

Note: The symbol "D" stands for durometer in Shore A units. UE stands for ultimate tensile elongation as a percentage of original length. Hacthane 110-M is Hacthane 110 plus 1.0% Irganox 1010 antioxidant.

18.3.3 OPTIMIZED ENCAPSULANT OR POTTANT FOR ADHESION

An earlier adhesion improvement study taught us that the epoxy resin Cardolite NC-547 added to Hacthane 110 improved its adhesion to PBT plastic (polybutylene terephthalate). PBT had proven to be the hardest substrate to bond to. In order to optimize adhesion even further, we undertook an extensive parametric study as described in the following table:

TABLE 18.3 Parametric Study of Lap Shear Strength under Varied Formulations and Conditions

Base formula	NC-547 added	Substrate	Condition	Duration
UH-510 A	0, 5, 10, and 15%	PBT	Ambient	Initial
UH-510 A	0, 5, 10, and 15%	Aluminum	125°C	2 Weeks
UH-510 A	0, 5, 10, and 15%	E-metal	85°C/95% RH	4 Weeks
Hacthane 110 A	0, 5, 10, and 15%	PBT	125°C	4 Weeks
Hacthane 110 A	0, 5, 10, and 15%	Aluminum	85°C/95% RH	Initial
Hacthane 110 A	0, 5, 10, and 15%	E-metal	Ambient	2 Weeks
Modified Hacthane 110A	0, 5, 10, and 15%	PBT	85°C/ 95% RH	2 Weeks
Modified Hacthane 110A	0, 5, 10, and 15%	Aluminum	125°C	4 Weeks
Modified Hacthane 110A	0, 5, 10, and 15%	E-metal	Ambient	Initial

Here is what we learned from that study: All of the lap shear strength data fell within a narrow range, from 156 to 402 psi. The three base formulations seemed very similar, one to another. Increasing amount of NC-547 seemed to improve lap shear strength overall. Adhesion to PBT was clearly improved by NC-547.

18.3.4 *OPTIMIZED REACTIVE POLYURETHANE HOT MELT ADHESIVE*

The best Hacthane compounds, formulated to be reactive hot melts, had the designations URHM-1, URHM-6, URHM-7, and URHM-8. They were all solids at room temperature, melted between 53 and 77°C, and could be applied to circuit boards at about 100°C using spray equipment. These hot melts proceeded to moisture cure over the next few days at room temperature becoming cross-linked rubbers in the process. The cross-linked coatings ranged in Shore D durometer from 38 to 46, in glass transition temperature from −25 to −9°C, had excellent adhesion, and no longer melted at 160°C.

Electrically, the series had initial insulation resistance in the 10^{12} to 10^{13} ohm range and proved to be hydrolytically stable in aging tests. They had low dielectric constants and dielectric strengths over 600 V/mil.

18.4 APPLICABLE PATENTS

TABLE 18.4 Applicable Hacthane Patents

U.S. Patent	Date granted	Description
5,185,498	February 9, 1993	Optimized potting compound for air bag sensors
5,385,966	January 31, 1994	Fillet-forming, thermal transfer adhesive
5,510,138	April 23, 1996	Reactive polyurethane hot melt conformal coatings
5,608,028	March 4, 1997	Optimized potting compound for air bag sensors

18.5 PUBLICATIONS

(1) "Novel, Thermally-Conductive, Elastomeric Adhesive for Electronic Components Assembly" by D.T. Chow and R.D. Hermansen was published as part of the SAMPE Electronic Materials Conference in Seattle in June 1988.

(2) I was asked to be chairman of a session at the International SAMPE Electronics M & P Conference held in New Jersey June 20–23, 1994. Henry M. Sanftleben (Hank) gave a paper on the air bag sensor potting development effort in my session.

KEYWORDS

- **Hacthane**
- **pottant**
- **encapsulant**

CHAPTER 19

TRANSPARENT POLYURETHANE PLASTICS

CONTENTS

19.1   WHAT ARE TRANSPARENT RIGID POLYURETHANES?

When one thinks of polyurethanes, they might think of flexible foam, rigid foam, elastomers, coatings, or even thermoplastic urethane rubber, but they are unlikely to think of thermosetting rigid plastics. It is a niche areas and within that niche is a subset of non-yellowing, water-white, transparent plastics. It is this particular niche and its technological development that is our focus here. The fact that they are non-yellowing is significant. Polyurethanes containing aromatic groups such as benzene rings comprise the majority of polyurethane usage. These materials will turn yellow to orange to brown with increasing sun exposure. However, the use of aliphatic isocyanates can prevent such discoloration.

19.2   A BRIEF RECAP OF THE TRANSPARENT POLYURETHANE EVOLUTION

John Raffo and I pursued the development of water-white, transparent, rigid polyurethane plastic during our own free time simply out of curiosity. Management decided it was a potential solution to the grounded F-15 fleet of planes. Studies were conducted to optimize the adhesion of the polyurethane cladding to the polycarbonate. "Fusion Bonding" and "Cast In Situ" techniques were both optimized. The impact properties of Sierraclad™ composites or monolithic clad specimen were extensively tested. The results were very promising. The deep forming of domes showed that the formulations were too cross-linked. More linear clads had been formulated and tested. Abrasion resistance and rain impingement resistance were evaluated. Finally, it was realized that weathering was a problem for the more linear clads and an extensive effort to improve this deficiency was undertaken via stabilizers.

19.3   MAJOR BREAKTHROUGHS

The following accomplishments were breakthroughs in the development of transparent polyurethane plastics: (1) bubble-free preparations, (2) extensive range of formulations, and (3) stabilization.

19.3.1 BUBBLE-FREE PREPARATIONS

If one were to simply combine a polyisocyanate and a short-chain polyol, the result would be a rigid foam. We found a technique early in our trails to remove absorbed water from the polyols. When we employed short-chain diols in our study, they held on to water more tenaciously. In order to remove dissolved water from the polyols, we heated the polyol in a vacuum flask to 10 or 20 degrees above the boiling point of water, while stirring the polyol with a magnetic stirrer and applying a vacuum to the flask. Polyols, based on a polyoxypropylene chain, tend to have limited heat stability and may benefit from the addition of an antioxidant. Secondly, we used only aliphatic isocyanates for two reasons: (1) to prevent yellowing and (2) to reduce the tendency of the isocyanate to react with water. These techniques allowed us to make water-white, bubble-free, transparent, polyurethane plastics.

19.3.2 EXTENSIVE RANGE OF FORMULATIONS

Thousands of non-yellowing, transparent, polyurethane plastic compounds, along with several of their properties could be compiled from our work on this project. Several experiments were conducted to improve

TABLE 19.1 Various Properties of the Clad 9 Series Transparent Plastics

Formula-tion	Shore D durometer at 7, 75, and 150°F	% L/T, % haze after salt abrasion, number of blasts	Tensile strength, psi	Tensile modulus, psi	Ultimate elongation, %
Clad 9-A	7°F: 75	0: 91.0/0.5	6120	260K	6.3
	75°F: 73	5: 90.6/4.7			
	150°F: 67	10: 88.7/9.8			
Clad 9-B	7°F: 75	0: 90.8/1.15: 90.2/5.8	6900	290K	29.0
	75°F: 73				
	150°F: 66	10: 87.2/15			
Clad 9-C	7°F: 75	0: 91.3/1.0	7720	300K	41.0
	75°F: 74	5: 89.4/7.9			
	150°F: 65	10: 87.1/19			
Clad 9-D	7°F: 76	0: 91.0/0.8	7980	300K	137.0
	75°F: 73	5: 86.5/14			
	150°F: 68	10: 83.6/34			

dome-forming ability. The following is typical: Clad 9 was made into a more linear polymer by increasing its diol content in steps. Both Teracol 1000 and di-propylene glycol were increased, going from Clad 9A to 9D. The Shore D durometer of all four clads was very similar from 7°F to 150°F. However, the change in light transmission (L/T) and haze after salt abrasion was markedly different from Clad 9A to 9D. The influence of polymer cross-link density on tensile properties is also marked. As linearity increased, strength and especially ultimate elongation vastly improved.

19.3.3   STABILIZATION

Extensive testing of the weathering characteristics of these transparent polyurethanes was conducted and we also learned which stabilizers are most effective in improving different kinds of polymer degradation. The combination of Eastman RMB UV absorber and Irganox 1010 antioxidant proved to be the best in slowing degradation and maintaining transparency. Accelerated testing was correlated with actual outdoor weathering results.

19.4   APPLICABLE PATENTS

U.S. 4,045,269 "Transparent formable polyurethane polycarbonate lamination" was granted to David L. Voss, William A. Miller, and Ralph D. Hermansen on August 30, 1977. The application had been filed on December 22, 1975.

KEYWORDS

- **transparent polyurethane**
- **transparent plastic**
- **stabilization**

PART VIII
Spin-Off Applications

CHAPTER 20

SPIN-OFF FROM OUR THERMAL TRANSFER ADHESIVES PATENTS

CONTENTS

This section is the forward-looking part of the book. Most of the patented inventions, which we have discussed, have expired patents or they soon will be expired. The technology is available for all to use. However, we need to think beyond the applications for which they were developed. I try to identify what was the breakthrough that made the invention patentable and project to where else that trait might be useful.

20.1 OVERVIEW

Keeping circuitry from overheating was a recurring problem in aerospace applications. It was especially true for space applications due to the absence of air. We were awarded five patents in this area of technology, which are listed in Table 20.1. Patents 4,866,108 and 5,385,966 resulted from our effort to replace polysulfide thermal transfer adhesives with non-outgassing equivalents. The new filleting adhesive became Hacthane 100, a polybutadiene-based polyurethane rubber cured with an aliphatic diisocyanate, which gave the uncured rubber a multi-month storage life as a frozen premix. Actually that accomplishment is the big breakthrough in my opinion. However, the Hacthane 100 paste could not be made to function as a frozen premix film adhesive, so we solve that problem by inventing Flexipoxy 100. We took the bold step of using flexible epoxy resins as main ingredients when the convention wisdom was to use them at about 15% level with DGEBA epoxy resins.

TABLE 20.1 Our Patents for Thermal Transfer Adhesives

Patent #	Title	Book location
4,866,108	Flexible Epoxy Adhesive Blend	Chapter 4
5,367,006	Superior Thermal Transfer Adhesive	Chapter 6
5,385,966	Frozen Premix, Fillet-Holding, Urethane Adhesives/Sealants	Chapter 3
6,060,539	Room-Temperature Stable, One-Component, Thermally-Conductive, Flexible Epoxy Adhesives	Chapter 11
6,132,850	Reworkable, Thermally-Conductive Adhesives for Electronic Assemblies	Chapter 10

20.2   FROZEN PREMIX, THERMALLY CONDUCTIVE, FLEXIBLE EPOXY ADHESIVES

20.2.1   FLEXIPOXY 208BF

Patent 5,367,006 was the result of a very extensive formulating effort to improve the dielectric properties of Flexipoxy 100 and extend its low-temperature service range. It was a secondary goal to be able to have the filleting adhesive and the film adhesive should be of the same formulation. There had been problems of Hacthane 100 and Flexipoxy 100 being incompatible if in contact with one another. It looked as though Flexipoxy 208BF met all of our goals. We got the patent and told our story to the world at a SAMPE convention only to learn that our NASA outgassing had been obtained erroneously. The culprit ingredient was Cardolite 547 epoxy resin, which was also the key ingredient that made success possible. Cardolite representatives could not help us solve the outgassing problem. The good news is that in the process of developing Flexipoxy 208BF, we developed a whole new technology with huge spin-off potential.

20.2.2   SPIN-OFF FROM FLEXIPOXY 208BF TECHNOLOGY

The flexible epoxy compounds resulting from Cardolite NC 547 resin and DP-3680 curing agent had remarkable properties: Durometers of 70 Shore A, volume resistivities greater than 10^{14} ohm-cm, glass transition temperatures below $-50°C$, and excellent processing temperatures. The one major problem precluding it for space applications was the very poor outgassing behavior of NC 547. However, there are numerous applications other than space where the compound would be very attractive. Here are some of them:

1. Encapsulation of terrestrial electronic devices.
2. Conformal coatings of printed wiring boards. Addition of some Cab-O-Sil to the formulation would give it better uniformity of thickness when coating components. A volatile solvent may be added to enhance flow.
3. Low-temperature elastomeric material applications (e.g., snowmobiles and other applications, where retention of flexibility to low

temperatures is important). NC-547/DP3680 flexible epoxies compete with polyurethanes for low-temperature flexibility, but have other advantages as well. For one, they do not foam when in contact with water. Substrates such as wood contain absorbed moisture, which will cause a polyurethane to foam. The Flexipoxies will not foam under the same circumstances.

20.3   ROOM TEMPERATURE-STABLE, ONE-COMPONENT, THERMALLY CONDUCTIVE, FLEXIBLE EPOXY ADHESIVES

Patent 6,060,539 demonstrated that flexible epoxies can be single-component, room temperature-stable systems with a low cure temperature acceptable for electronic assembly. Extensive testing for specific applications is mandatory before using them though. We were only interested in showing that they were feasible. The supplier could keep them in freezer storage until ready to ship in order to extend storage life.

20.4   ATTAINING THERMAL CONDUCTIVITY BY IMPREGNATING THE FILLER

Patent 6,132,850 represents a different method of applying a thermal transfer adhesive. Higher thermal conductance is attained if we impregnate alumina particles tightly packed into the bondline void. Flexible epoxies were formulated to be low viscosity impregnants, be dielectric, and easily reworkable. These formulations may function well for future impregnation applications or serve as starting point formulations to be modified for specific properties.

20.5   FROZEN PREMIX POLYURETHANES

One of the big bonuses that came from the development of Hacthane 100 adhesive was that we now had an elastomeric adhesive which can be used as a one-component system. It is important to avoid weighing and mixing on the assembly line. What is unique about our invention was finding a way to make it into a one-component frozen premix compound having months of storage life. We manufactured Hacthane 100 adhesive for many years and learned how to get as much as 9 months of storage life at −40°C. The key is to eliminate absorbed water in the filler and the polyol. The frozen premix concept could be extended beyond thermal transfer adhesive to applications such as conformal coatings, encapsulants, casting compounds, molding compounds, sealants, and others.

Moreover, this adhesive has a glass transition less than −60°C. This very low temperature transition opens the use of this compound to applications where low operating temperatures are required. Using the basic formulation of polybutadiene polyol plus aliphatic diisocyanate, low temperature-performing adhesives, sealants, potting compounds, encapsulants, conformal coatings, and other uses are easily formulated.

That polybutadiene-based polyurethanes had superior dielectric properties has been known for decades. The frozen premix concept could be extended to other polyol types besides polybutadiene polyols. For example, polyether polyols and polyester polyols are widely available. The key is to find the least reactive di- or tri-isocyanate or have one synthesized. Aliphatic isocyanates are less reactive than aromatic isocyanates, but steric hindrance around the isocyanate group can be utilized to reduce the reactivity even more. Aliphatic isocyanates are more expensive than their aromatic counterparts, but in the case of aerospace or medical applications, price is a secondary consideration.

KEYWORDS

- **thermal transfer adhesives**
- **Flexipoxy**
- **NC 547**

CHAPTER 21

SPIN-OFF FROM OUR ELECTRICALLY CONDUCTIVE ADHESIVES PATENTS

CONTENTS

This chapter is concerned with electrically conductive adhesives. There were two projects, which led to patentable inventions: one was the NCMS-sponsored project to develop an organic solder and the second was our development of a room temperature-stable electrically conductive adhesive. Table 21.1 shows the location of the development story in the book.

TABLE 21.1   Our Patents for Electrically Conductive Adhesives

Patent #	Title	Book location
5,575,956	Room-Temperature Stable, One-Component, Electrically Conductive, Flexible Epoxy Adhesives	Chapter 11
5,929,141	Adhesive of Epoxy Resin, Amine-Terminated BAN, and Conductive Filler	Chapter 15

There were already formulations on the books at HAC for silver-filled rigid epoxy adhesives. It was a simple matter of adding DuPont's V-9 silver powder to an epoxy mixture in sufficient quantity to get electrical conductivity. However, our involvement in the NCMS project to qualify an organic solder, which is another name for an electrically conductive adhesive, took us in the direction of a silver-filled flexible epoxy formulation. The driver for this approach was drop resistance. One of the most difficult of the consortium's requirements was that bonded specimens pass the following drop test requirement of surviving six drops:

The drop test:

Conductive adhesive layer is 5–7 mils. Test specimen is a 44 pin, leaded PLCC bonded to an epoxy-glass rectangular board. Three specimens are individually dropped from 60 inch height vertically down a guide so that they hit on their edge. The number of drops survived is recorded.

Patent 5,929,141 represents the successful formulation work of many months effort toward the NCMS goals. The following table sums up our two patentable ventures into the area of electrically conductive adhesives.

21.1   THE NCMS FINALIST NCMS 7430 ELECTRICALLY CONDUCTIVE ADHESIVE

21.1.1   WE HAD THE ONLY DROP-RESISTANT, ORGANIC SOLDER

The electrically conductive adhesive NCMS 7430 is the result of a very extensive formulation development effort. Chapter 15 details the progression of formulations and their properties. The biggest challenge was meeting the drop resistance requirement. The early part of the NCMS effort involved evaluation of 25 commercially available electrically conductive adhesives. All of them failed to even come close to meeting the drop resistance requirement.

21.1.2   FLEXIPOXY TECHNOLOGY MADE IT POSSIBLE

We were more successful because we used flexible epoxy polymer as the binding agent for the silver flake particles. It was not certain that this approach would work. Flexible polymeric materials have a higher thermal expansion coefficient than rigid polymeric materials, which means the gap between particles might open with a small change in temperature. Electrical conductivity might suffer from that phenomenon. On the other hand, there only being a miniscule amount of polymer between silver particles also means that there is little volume of polymer in which polymer chains can deform to help dampen the shock wave from the drop.

21.1.3   A UNIQUE SILVER FILLER SAVED THE DAY

The difference in the geometry of the different vendor silver particles was the key to making meaningful headway. Apparently, there is an optimum silver particle size and shape for each distinct flexible epoxy polymer. Therefore, if you are formulating to meet a new set of property requirements, a practical approach would be to first optimize the flexible epoxy or other rubbery polymer adhesive, and second to find the optimum silver particle for that polymeric adhesive.

21.1.4   BUILD ON OUR DISCOVERY

In the world of formulating, we have to be cautious of spending research money on an effort that seems impossible. It is very valuable that someone has already done it. Even if you fail, you can claim that it was not impossible. Now the world knows that a drop-resistant organic solder is possible! That is a big_deal because mobile devices often get dropped and they are a major part of electronics these days.

More than knowing that it can be done, we have shown you step-by-step how we did it. You do not have to reinvent the wheel.

21.2   ROOM TEMPERATURE-STABLE, ELECTRICALLY CONDUCTIVE ADHESIVES

Patent 5,575,956 covers the technology behind formulating a room temperature-stable, one-component, electrically conductive flexible epoxy adhesive. Although the feasibility of such an adhesive was proven, minimal characterization was done. Anyone using this technology should verify that the compound has the thermal stability, hydrolytic stability, and other application tests as are necessary before using these particular formulations.

KEYWORDS

- electrically conductive adhesives
- NCMS
- silver particle

CHAPTER 22

SPIN-OFF FROM OUR ENCAPSULANTS, POTTING COMPOUNDS, AND IMPREGNANTS PATENTS

CONTENTS

Many applications in electronic assembly involve filling a volumetric space with a dielectric polymer to protect the circuitry inside that space. These polymeric materials are called encapsulants if they can stand alone and potting compounds if they require a permanent can or container for exterior protection. Many people use the terms "encapsulant" or "potting compound" interchangeably. Sometimes, the volumetric space is first filled with particulate matter such as sand, glass beads, tabular alumina, etc. Then, a low viscosity polymeric liquid impregnates the filler particles by penetrating the void areas between filler particles. This polymeric liquid is called an impregnant. However, once the polymeric liquid cures, the composite material is actually an encapsulant or potting compound.

The patents of this book, which fit into the category of encapsulant, potting compound, or impregnant, are listed in Table 22.1.

TABLE 22.1   Our Patents for Encapsulants, Potting Compounds, and Impregnants

Patent #	Title	Book location
5,185,498	Circuit Assembly Encapsulated with Butadiene Urethane	Chapter 12
5,350,779	Low Exotherm, Low Temperature Curing, Epoxy Impregnants	Chapter 8
5,457,165	Encapsulant of Amine-Cured Epoxy Resin Blends	Chapter 12
5,608,028	Polybutadiene Urethane Potting Material	Chapter 12
6,723,803	Adhesive of Flexible Epoxy Resin and Latent Dihydrazide	Chapter 11
6,132,850	Reworkable, Thermally-Conductive Adhesives for Electronic Assemblies	Chapter 10

Note: Alternative use for 6,723,803 and 6,132,580.

These compounds can be further categorized by their type of polymer. Our polyurethane formulation has been given Hacthane names. The flexible epoxies are called Flexipoxy. The spin-off possibilities are discussed in the sections below.

22.1   HACTHANE ENCAPSULANTS AND POTTING COMPOUNDS

22.1.1   HACTHANE 115, AN OPTIMIZED POTTING COMPOUND

The applicable patents are 5,185,498 and 5,608,028, both of which resulted from our technical assistance to Delco Electronics on the potting compound for an air bag sensor. Both of these patents have expired, so this technology is in the public domain.

Hacthane 110 was our first formulation match for PC-1234, which was the best candidate to comply with Delco Electronics' rigorous require-ments. It is composed mainly of a hydroxyl-terminated polybutadiene liq-uid polymer, and a liquid version of methylene diisocyanate. These two reactants would produce a polyurethane quite soft with excellent electrical properties. There was also a third ingredient, a polyoxypropylene triol, included in the formulation to raise the hardness, physical properties, and cross-linking.

Hacthane 115 is the end product of a very extensive optimization pro-gram to enhance key properties of Hacthane 110. Hacthane 115 is superior to Hacthane 110 in two major ways: (1) the best antioxidant at its optimum concentration is in the Hacthane 115 formulation. This modification helps preserve the flexibility of the potting compound over long durations in hot environments. Otherwise, the compound becomes too hard and brittle. (2) The adhesion of the potting compound to its container has been optimized by incorporating Cardolite NC-547 into the formulation.

22.1.2 THERE ARE NUMEROUS SPIN-OFF APPLICATIONS FOR HACTHANE 115

Hacthane 115 potting compound was developed to survive in the very hostile environment of an automobile engine compartment. Autos are designed to function in all of the climates of the world, so this potting compound can function anywhere in the world. Moreover, chemicals such as gasoline, oil, hydraulic fluid, salt, and others did not erode the pot-ting material. The range of applications could be extended from autos to motorcycles, snowmobiles, watercraft, road-building equipment, mining vehicles, industrial machinery, and others.

OTHER APPLICATIONS INCLUDE:

1. Use as an encapsulant: If Hacthane 115 were slightly harder, it would serve as an encapsulant too. Either an increase in the triol ingredient plus corresponding increase in isocyanate or use of cas-tor oil in place of the triol would accomplish this task.

2. Transparent potting compound: Elimination of the colorant in Hacthane 110 would result in a transparent potting compound. This would be desirable if the components needed to be inspected. Elimination of the triol from the formulation and possibly the addition of mineral oil as a plasticizer would make this compound easily reentered for circuitry repair. Use of an aliphatic diisocyanate, such as Hylene W in place of Isonate 143L, would improve transparency too.

3. Hacthane 115 potting compound remains flexible down to very low temperatures. Its glass transition temperature is about $-70°C$. Therefore, applications requiring low temperature flexibility give Hacthane 115 an advantage. Arctic or Antarctic applications as well as planetary probes come to mind.

4. Packaged as a frozen premix compound: A version of Hacthane 115 could be formulated so that it could function as a one-component frozen premix. Substituting an aliphatic diisocyanate for the Isonate 143L will do the trick. Hylene W, isophorone diisocyanate(IPDI), and others are suitable candidates.

5. Offered as an adhesive: Considerable experimental effort was expended to assure that Hacthane 115 adhered well to various substrates and maintained adhesion level under dry heat or humid heat aging conditions. Hacthane 115 can be useful as an elastomeric adhesive where reliability is important.

22.2   FLEXIPOXY ENCAPSULANTS AND POTTING COMPOUNDS

22.2.1 FLEXIBLE EPOXIES CAN MAKE GOOD ENCAPSULANTS AND POTTING COMPOUNDS

Patent 5,457,165 was awarded for our development of flexible epoxy encapsulants. The work was funded by Delco Electronics, although it employed formulating concepts that we had learned through solving Hughes Aircraft Company's problems. The airbag sensor potting compound had to survive severe operating conditions, including high and low temperatures, high and low humidities, and contact with salt spray, gas, oil, and hydraulic fluid. They wanted a rubbery rather than rigid polymer, based on their extensive testing of submitted candidates.

Flexipoxy 313 and 315 survived rigorous screening tests, which included a dry heat exposure test and a heat plus humidity exposure. The dry heat exposure was for 7 days at 125°C. The heat plus humidity exposure was 7 days at 85°C plus 95% relative humidity.

22.2.2 SPIN-OFF APPLICATIONS

A few examples of possible new applications from the Flexipoxy technology are listed below:

1. Frozen premix encapsulants: Whereas Delco Electronics was interested in these compounds packaged as two-component systems to be used in mixing/metering machines, they could also be packaged as frozen premix one-component systems and would have months of storage life at freezer temperatures.
2. Low-temperature encapsulants: Flexipoxy 208BF was developed to be a thermal transfer adhesive, but the unfilled version would remain flexible down to quite low temperatures.
3. Reenterable encapsulants: Encapsulation of buried telephone cable splices might also be an application for formulations derived from Cardolite NC547 cured with DP-3680.

22.2.3 FORMULATING FOR CUSTOM APPLICATIONS

By using the numerous examples of customizing flexible epoxies discussed in this book to meet a list of property requirements, one should be able to zero in on a custom formulation fairly rapidly. For example, the use of fillers in the formulation may be a way to increase thermal conductivity, reduce thermal expansion, reduce exotherm temperature, or reduce cost. All of the lessons learned in the formulation of thermal transfer adhesives apply to encapsulants as well. Cardolite NC-547 epoxy resin was effective in lowering the glass transition temperature while also improving the dielectric properties.

For unfilled formulations, Flexipoxy 313 is a good starting point. It had the best properties of all of those tested. Flexipoxy 315 might be a good starting point for filled formulations. It seemed too exothermic unfilled, but addition of mineral fillers helped to reduce exotherm. Also, Flexipoxy

315 would become less exothermic if the Heloxy 67 resin is blended with other higher-molecular-weight resins.

22.3   FLEXIPOXY IMPREGNANTS

22.3.1   FLEXIBLE EPOXIES CAN MAKE GOOD IMPREGNANTS

Flexible epoxies are uniquely suited for use as impregnants due to their versatility; they can easily be formulated to have the properties required for different applications. In the application which led to patent 6,132,850, titled "Reworkable, Thermally-Conductive Adhesives," there was a need for an impregnant which could impregnate alumina particles and bond them together, have good dielectric properties, and have a controlled adhesive strength. The adhesive strength should be strong enough to hold the assembly of circuitry, heat sink and filler particle together, yet weak enough to remove the heat sink without damaging the circuitry or components.

Three formulations were cited in the patent as having the properties mentioned above. Two of them had better compliance:

TABLE 22.2   Flexipoxy RDI-100 and RDI-101, Reworkable, Dielectric, Impregnant/ Adhesives

Flexipoxy number	Formulation	Shore A, initial	Shore A, heat aged	Viscosity, initial, cps	Viscosity, 1 h later
RDI-100	Heloxy 505 100 pbw DP-3680 28 pbw	20	33	600	900
RDI-101	Heloxy 505 100 pbw Dytek A 4.8 pbw	18	32	550	800

The very low Shore A durometer is indicative of reworkability. The slow rise in viscosity is indicative of viability as an impregnant. There is adequate time for the impregnant to penetrate the packed filler particles before it becomes too viscous due to advancing polymerization.

22.2.3   POTENTIAL SPIN-OFF

In addition to the alumina particles considered in the application above, many other filler particles might be impregnated. Sand, for example, is a low-cost filler. Spherical particles, like glass beads, are especially easy to impregnate. Platelet or rod-shaped particles may be difficult to penetrate without leaving voids. These high aspect ratio fillers also tend to raise the cohesive strength of the cured polymer. Layers of cloth and other porous materials might also be impregnated.

Numerous Flexipoxy formulations have been discussed in this book along with many of their properties. The unfilled formulations can be considered for impregnation applications. There are numerous factors to consider when choosing a Flexipoxy impregnant. To how low a temperature must the polymer be rubbery? What upper temperatures might be expected? How difficult is the impregnation? What adhesive strength level is expected? Should it be reworkable?

If the polymer is expected to be weak for easy disassembly, then low durometer flexible epoxies should be considered. Plasticizers may also be added to the formulation to weaken the cohesive strength and lower viscosity of the liquid mixture. If a tough rubbery matrix is desired, higher durometer flexible epoxies should be considered.

22.4   RIGID EPOXY IMPREGNANTS

22.4.1   CONTROLLING THE CURE TEMPERATURE RANGE VIA FORMULATING TRICKS

Patent 5,350,779 describes a technology for formulating rigid epoxy impregnants which cure at lower temperatures than the currently used impregnant. The problem arose from high-voltage power supplies which failed due to the residual stresses in the cured composite encapsulant. The composite consists of tightly packed alumina particles, which have been impregnated by the epoxy impregnant and oven cured. The new lower-temperature-curing impregnants are possible due to a blend of unhindered and hindered diamine curatives.

Another important consideration is the danger of a runaway exotherm event due to the introduction of the unhindered diamine curative. The use of an unhindered diamine curative makes lower cure temperatures possible. However, its exclusive use may be dangerous. The compromise is to blend hindered and unhindered diamines together to achieve safe exotherms and lowered cure temperature.

In the case of the SAP 391 Electronic Power Conditioner (EPC) problem, menthane diamine, a hindered cycloaliphatic diamine, was the sole curative. It had to be cured at temperatures of 160°F or higher to achieve an adequate cure. However, residual stresses contributed to high-voltage breakdown of the dielectric. Stress analysts calculated that a cure temperature of 130°F or lower would adequately lower these stresses. We made this possible by blending PACM, an unhindered cycloaliphatic diamine, with menthane diamine to achieve a safe, lower temperature-curing impregnant possible.

22.4.2 SPIN-OFF FROM NEW DIAMINE BLENDS

A host of other blends are possible to impregnant formulators. For example, the following table presents a list of hindered and unhindered cycloaliphatic diamines, which can be blended to ideal curative systems.

TABLE 22.3 Cycloaliphatic Curatives for Epoxy Impregnants

Sterically hindered (yes/no)	Curative name	Chemical name
Yes	MD	Menthane diamine
Yes	IPDA	Isophorone diamine
Yes	Lamiron C-260	Para-bis (amino-meta-methyl-cyclohexyl) methane
No	PACM	para-bis (amino-cyclohexyl) methane
No	MXDA	Meta-xylylene diamine (see note 1)
No	DACH	1,3-diamino-cyclohexane

Note 1: Although MXDA is an aromatic diamine, its amine groups are attached to methylene groups rather than to the benzene ring and behave somewhat like cycloaliphatic diamines.

Of the three hindered cycloaliphatic diamines in the table, menthane diamine (MD) is the most hindered and most unreactive. Both of its amine groups are surrounded by bulky groups, which reduce the probability that a collision with an epoxide group will be favorable for reaction. IPDA, by comparison, has one highly hindered amine group, but the other amine group is unhindered. In the case of Lamiron C-260, both amine groups are only hindered on one side, and would be more reactive than MD. So in the case of IPDA or Lamiron C-260, lower minimum cure temperatures than we attained with MD can be expected. Cost and availability may determine which curatives are best to use.

22.4.3   SPIN-OFF VIA EPOXY RESIN MODIFICATION

Within the patent claims, we discussed different ways of attaining a low-viscosity resin component. In the application cited, Epon 815 was resin component with curative menthane diamine. Epon 815 is actually 89% Epon 828 and 11% butyl glycidyl ether (BGE), which is a monoepoxide. Epon 815 is a good choice for the epoxy resin of an impregnant, because it is low in viscosity. However, the formulator should understand one fact about monoepoxides. When a monoepoxide BGE molecule reacts with one of the four active hydrogens on the diamine curative, it creates a terminal point. It also reduces the functionality of the curative from four to three. In other words, the cured epoxy polymer will be less cross-linked and its properties will be slightly different as a result. There is another way to formulate the impregnant. A low-viscosity diepoxide could be used instead of the monoepoxide. We examined Heloxy 67 diepoxide in place of BGE. It successfully lowered the viscosity of the resin blend.

The formulator may be attempting to develop an impregnant to match a list of property requirements, such as a maximum viscosity increase versus time, a minimum tensile strength, or a minimum volume resistivity. The selection of monoepoxide or low-viscosity diepoxide in the resin blend may make the difference in meeting the requirement.

22.5   ROOM TEMPERATURE-STABLE, ONE-COMPONENT ENCAPSULANTS AND POTTING COMPOUNDS

Patent 6,723,803 covers room temperature-stable, one-component flexible epoxy adhesives. These formulations might function as encapsulants, potting compounds, or impregnants just as easily. Our patent only proved the feasibility of such compounds. Application testing should precede using them in any application.

KEYWORDS

- **Hacthane 115**
- **encapsulant**
- **impregnants**

CHAPTER 23

SPIN-OFF FROM OUR HIGH-VOLUME ELECTRONIC ASSEMBLY MATERIALS PATENTS

CONTENTS

In this section, we will consider the spin-off potential of the materials in Table 23.1. These electronic assembly materials were developed for very high-volume implementation in automotive electronics manufacturing.

TABLE 23.1   Our Patents for Innovative Materials for High-Capacity Manufacturing

Patent #	Title	Book location
5,708,056	Hot Melt Epoxy Encapsulation Material	Chapter 14
5,759,730	Solder Joint Lead Encapsulant	Chapter 14
5,965,673	Epoxy-Terminated Prepolymer of Polyepoxide and Diamine with Curing Agent	Chapter 12
5,510,138	Hot Melt Conformal Coating Materials	Chapter 13

23.1   REACTIVE HOT MELT CONFORMAL COATINGS

23.1.1   RUBBERY REACTIVE HOT MELTS

One of the most novel aspects of our patent 5,510,138 is that we were able to formulate rubbery reactive epoxy hot melt compounds for the first time. Although the focus was on developing reactive hot melt conformal coatings, it was recognized that the same technology could apply to adhesives, encapsulants, and other uses. The basic concept was to apply a hot melt conformal coating to circuit boards and eliminate the usual post-processing such as oven cures, UV-lamp curing, and so forth. Instead, the hot melt would quickly solidify upon cooling and the boards would advance into the next stage of processing. However, we realized that the thermoplastic state of a hot melt was an unsatisfactory condition for the conformal coating because it would liquefy at higher temperatures and be susceptible to certain solvents. Thus, a mechanism was needed to cross-link the applied conformal coating and convert it into a non-melting, insoluble, thermoset.

Patent 5,510,138 reveals how we accomplished that feat for flexible epoxy hot melt compounds. Actually, two different mechanisms were shown to do the job. One approach was to use the ingress of moisture from the air to activate cross-linking reactions in the RHMCC (i.e., reactive hot melt conformal coating). The second mechanism is to use elevated temperatures experienced in post-operations activate cross-linking reactions in the RHMCC. The trial formulations in our patent and their properties are shown in Table 23.2.

TABLE 23.2 Reactive Hot Melt Compounds Using Heat or Moisture Activation to Cross-Link

Formulation name	Cross-linking activation mechanism	Melting point, °C	Durometer, Shore A
RHM-L1	Heat	39	82
RHM-L2	Heat	45	45
RHM-L3	Heat	41	75
RHM-L4	Heat	40	36
RHM-L5	Heat	40	40
RHM-K1	Moisture	39	82
RHM-K2	Moisture	39	82
RHM-K3	Moisture	39	82
RHM-K4	Moisture	41	82
RHM-K5	Moisture	40	85
RHM-K6	Moisture	40	80

Notice that the melting points of all the compounds are about 40°C, which is a very safe temperature for the electronic components. The Shore A durometers are typical of rubbery materials, which is a necessity for a conformal coating.

23.1.2 MAKING OUR OWN EPOXY PREPOLYMERS

It is relatively easy to formulate a rigid epoxy hot melt compound because numerous rigid epoxy resins are available in solid form. However, we needed to formulate a rubbery hot melt material, and no flexible epoxy resins were found in our industry search at that time. In order to create a solid flexible epoxy resin, we experimented with making epoxy prepolymers. Essentially, the process involves reaction of diepoxides with a diamine. There are four active hydrogens in a diamine molecule, and one epoxide group will add to one active hydrogen site on the diamine. So, four diepoxide molecules added to one diamine molecule should yield a new molecule having four free epoxide groups available. If either the diepoxides or the diamine has a flexible backbone, the prepolymer will contain a flexible backbone. And now the final test, is the new prepolymer a solid with a melting point in our temperature range of interest? If yes, mission accomplished.

Patent 5,965,673, epoxy-terminated prepolymer of polyepoxide and diamine with curing agent, covers our technology in preparing solid, rubbery, epoxy prepolymers. Within the patent, we describe several epoxy-terminated prepolymers being prepared by reacting liquid diepoxides with a 4000 molecular weight solid polyether diamine.

23.1.3   SPIN-OFF FROM PATENT 5,965,673

I recently examined the 2014 Huntsman product bulletin for Jeffamine polyether amines and noticed several other diamines which might yield interesting candidates for reactive hot melt development. They are:

1. Jeffamine D-4000 (XTJ-510) has a polyoxypropylene chain of about 68 repeating units between terminal primary amines. It is a liquid.
2. Jeffamine XTJ-542 has a poly(oxy-tetra-methylene) chain between terminal primary amines. It is a solid below 9°C.
3. Jeffamine XTJ-548 has a poly(oxy-tetra-methylene) chain between terminal primary amines. It is a solid below 33°C.
4. Jeffamine XTJ-542 has a poly(oxy-tetra-methylene) chain between terminal primary amines. It is a solid below 16°C.

23.1.4   SPIN-OFF FROM PATENT 5,510,138

Reactive hot melt conformal coatings have advantages over traditional conformal coatings because it is not necessary to provide ovens, UV lamps, for the curing of the conformal coating and because the coated board can go immediately into the next operation. The patent provides detailed information on how to produce such a product.

The moisture-curing reactive hot melts are only applicable to thin cross-sections like a coating. However, the heat-activated reactive hot melts could be used for a variety of applications. Adhesives, potting compounds, encapsulants, casting compounds, and sealants are a few of them.

23.2   LOW-TEMPERATURE CURING, SOLDER JOINT LEAD ENCAPSULANTS

23.2.1   REACTIVE HOT MELT SJLE DESCRIPTION

Solder joint lead encapsulants (SJLE) keep soldered leads from debonding. Electronic circuit assemblies often must survive under hostile operating conditions. Such assemblies often employ surface mount integrated circuit packages, which attach to the printed wiring substrate by soldering their leads. Outdoor applications typical for autos or military equipment may see many hundreds of thermal cycles which apply stress to the soldered leads. Such stressing can cause fatigue and fracturing of the soldered leads. The expansion of conformal coatings, under the leads, makes the problem worse by putting a tensile stress on the soldered joint.

Delco Electronics engineers required A SJLE which cured at lower temperatures than the 150°C required by commercially available SJLEs. Our approach was to develop reactive hot melt rigid epoxy compounds, which could be applied at moderate temperatures (45–90°C) and then cross-linked during subsequent operations. Patents 5,708,056 and 5,759,730 cover essentially the same reactive hot melt SJLE invention. I believe that separate patents exist to give patent rights to both Delco Electronics and Hughes Aircraft companies.

Four SJLE formulations were presented, which prove the concept. All four had the following properties:

(1) Melting points in the hot melt stage of about 35°C.
(2) A viscosity of 1000 cps at a temperature between 45 and 90°C. for hot melt application.
(3) A glass transition temperature greater than 115°C when fully cured to a thermoset.
(4) Coefficient of thermal expansion of about 30–34 ppm/°C when fully cured to better match the leads and solder.
(5) Cross-linking to a thermoset under mild conditions, such as 30 min at 120°C.

23.2.2 *SPIN-OFF FOR OUR LOW-TEMPERATURE CURING SJLES*

Customized SJLEs to match different manufacturing conditions are possible using the formulating technology discussed in these patents.

The concept of an adhesive, encapsulant, sealant, or coating which can be applied as a hot melt material for rapid processing and which converts to an insoluble, non-melting, rigid thermoset during subsequent thermal excursions applies to a multitude of applications. For example, film adhesives could be made from our invention which bond substrates together very quickly for volume production. It is a matter of melting the hot melt film adhesive, applying pressure, cooling the assembly, and setting the bonded part aside. The latent curative gets activated during exposures to higher temperature, converting the adhesive to a thermoset.

The SJLEs of our patents had glass beads in the formulations to match the coefficient of thermal expansion of the SJLE to that of solder. New applications may make the glass beads unnecessary or may require a different filler. For example, aluminum powder as a filler would make the hot melt conduct heat or cool more readily, which could speed up processing times.

KEYWORDS

- hot melt
- SJLE
- epoxy prepolymer

CHAPTER 24

SPIN-OFF FROM OUR SPECIALTY ADHESIVES PATENTS

CONTENTS

Two of our inventions were specialty adhesives. In other words, these adhesives had to be specially formulated for their applications because no such adhesives existed as far as we knew. Table 24.1 lists the patent numbers, patent titles, and where the story of their invention can be found in this book.

TABLE 24.1 Our Patents for Specialty Adhesives

Patent #	Title	Book location
4,940,633	Method of Bonding Metals with a Radio-opaque Adhesive/Sealant for Void Detection and Product Made	Chapter 7
5,780,581	Platable Structural Adhesive for Cyanate Ester Composites	Chapter 9
5,840,829	Platable Structural Adhesive for Cyanate Ester Composites	Chapter 9

24.1 RADIOPAQUE ADHESIVE/SEALANT

24.1.1 USING HIGH Z FILLERS TO MAKE AN ADHESIVE RADIOPAQUE

This patent was granted in 1990 and has since expired, so the technology is now in the public domain. Our original engineering problem was an inability to conduct non-destructible tests for voids on an epoxy adhesive/sealant sandwiched between sheets of aluminum. The adhesive was not visible during x-ray inspection because the aluminum sheets were more radiodense than the adhesive. The device would contain a gas within it, so having a gas-tight seal was essential. We needed a way to assure ourselves that each device was free of voids and had that gas-tight seal.

The solution to the problem was to modify the adhesive so as to make it radiopaque. Theory had taught us that the effective Z number of the modified adhesive had to be much higher than the Z number of aluminum, which is 13. Tungsten powder was selected as an additive to the adhesive due to its very high Z of 74. Indeed, the addition of a tungsten filler to the adhesive did make the adhesive more radiodense than the aluminum to x-rays. Thus, it was possible to detect voids in the modified adhesive using x-ray inspection.

We also wished to retain all of the properties of the original adhesive if possible, so we sought to keep the tungsten powder addition to the lowest level possible. This was accomplished by preparing several modified versions of the tungsten-filled adhesive, where the Tungsten powder was incrementally increased. Then we prepared specimens for x-ray inspection similar in configuration to the actual part using the modified adhesives. It was soon obvious from the x-ray photographs which was the minimum level of tungsten powder that worked to make the adhesive visible for void detection.

24.1.2 SPIN-OFF

Spin-off from this special case to a myriad of different applications is quite simple to accomplish. For almost any adhesive or sealant sandwiched between almost any metal, a similar solution may be possible. Tungsten is a very high Z material, which gives it a superior advantage, but other high Z materials will work too. One might consult a periodic chart of the elements for ideas on alternative high Z fillers. Generally speaking, powdered fillers do not negatively affect the cure or properties of polymeric compounds, but the first step should be to verify that it is true for the adhesive or sealant you wish to modify. Then it is easy to determine how much high Z filler must be added to the adhesive or sealant in order to make it more radiodense than the metals in question. Prepare test specimens by duplicating the metals, thicknesses, and configurations of the application. Use high Z-modified versions of the adhesive or sealant in the test specimens. Subject the specimens to x-ray inspection and determine the level of high Z filler necessary to make the adhesive/sealant visible.

As an example of a different application, we examined steel instead of aluminum. Steel is mainly iron and iron has a Z of 26. So steel is approximately twice as radiodense as aluminum. Chromium and nickel have Zs close to that of iron. Experimentation showed that using the same adhesive (Epiphen 825A) and having a filler volume of 33% made the adhesive visible behind 0.10 inch thickness of steel. The filler was 50 weight % mica and 50 weight % tungsten.

Future applications of the radiopaque adhesive/sealant technology could include metal-bonded structures for industrial applications (chemical processing, food processing, robotic systems, etc.), marine applications (ships, submersibles, offshore oil rigs, etc.), military applications (tanks, aircraft, helicopters, drones, missiles, etc.), automotive applications(cars, buses, trucks, etc.), rail applications (locomotives, railcars, etc.) and space applications (satellites, orbiting laboratories, etc.)

24.2   PLATABLE STRUCTURAL ADHESIVE FOR CYANATE ESTER COMPOSITES

24.2.1   MAKING AN ADHESIVE SEEM LIKE THE CYANATE ESTER COMPOSITE THAT IT BONDS TOGETHER

The application was a new, lighter weight, microwave channeling device for an earth-orbiting satellite. A room temperature-curing adhesive was needed to bond composite units together. The units were composed of graphite fiber-reinforced cyanate ester composite. The sought adhesive would be amenable to the same etching and metal plating processes that had been optimized for the composite units. Ideally, the bonded assembly of units could be plated with one continuous coating of metal.

Hundreds of different adhesives were evaluated, but failed to meet the requirements. The breakthrough came when we realized that the adhesive needed to be composed mainly of the same material as the composite units, namely of cured cyanate ester plastic. If we could grind it into fine particles, it could be added as an inert filler to a room temperature curing, epoxy adhesive. It was not easy to prepare cyanate ester powder. Cryogenic temperatures were needed to embrittle the tough plastic.

Two patents were granted for this invention, 5,780,581 and U.S. 5,840,829. The latter patent supersedes the first patent. Within the patent are numerous adhesive formulations which employ the concept of a powdery cyanate ester filler in an epoxy adhesive and which yielded excellent plating results. There was one complication though. Many of the epoxy adhesives had a tacky surface if that surface faced the air. The degree of tackiness varied with epoxy formulation and with cure conditions. An experimental matrix of differing conditions is presented along with the results (tackiness, graininess, and plating adhesion). The best adhesive was 150 parts Epon 815, 18 parts TEPA curative, 50 parts of 25 micron particle size, and BtCy-1 cyanate ester powder. It was cured for 20 days against Ef-179 coated steel. It was not tacky, had the least graininess, and the best plating adhesion.

24.2.2 SPIN-OFF

There may not be a great deal of spin-off from this invention. Our application was considerable weight savings on a satellite where it is worth $50,000 or more to save one pound. We could afford an expensive solution to the problem. There may be application to reflective optics and other microwave channeling devices. I think the important lesson learned from this effort is that adding ground up powder from the plastic you wish to plate on to the adhesive works. Almost anything can be ground into a powder. Sometimes, you can utilize that fact to solve a problem.

KEYWORDS

- specialty adhesive
- radiopaque
- high Z filler
- cyanate ester composite

CHAPTER 25

SPIN-OFF FROM OUR SPECIALTY RIGID THERMOSETS PATENTS

CONTENTS

Most of the patents in this book were custom-formulated rubbery compounds. Only a few were rigid plastics. So our final category becomes "Specialty Rigid Thermosets." Of the three patents falling into this category, two of them have already been discussed as to possible spin-off. There is little more to say about Patent 5,350,779 or Patent 5,759,730. That leaves Patent 4,045,269 as our current and final focus.

TABLE 25.1   Our Patents for Specialty Rigid Thermosets

Patent #	Title	Book location
4,045,269	Transparent Formable Polyurethane Polycarbonate Lamination	Chapter 16
5,350,779	Low Exotherm, Low Temperature Curing, Epoxy Impregnants	Chapter 8
5,759,730	Solder Joint Lead Encapsulant	Chapter 14

25.1   TRANSPARENT, FORMABLE, POLYURETHANE, POLYCARBONATE LAMINATION

25.1.1   A NEW FAMILY OF TRANSPARENT, POLYURETHANE PLASTICS WAS INTRODUCED

The engineering problem prompting our invention was the canopies for the F-15 aircraft. Changing the canopies from acrylic to polycarbonate solved the deadly hazard of bird strikes at low altitude and high speed. A thin protective coating on the canopies solved the abrasion problem where polycarbonate lost its transparency. It solved the problem until the F-15s were flown in the rain, then the coating was lost. Our attempt to solve the problem utilized rigid, non-yellowing polyurethane (PU) cladding over the polycarbonate.

Numerous formulations of rigid, non-yellowing polyurethane were examined for optical clarity, impact resistance, deep dome forming, and weathering. Chapter 16 of this book tells the entire story and recounts the technology that we developed.

A typical formulation is presented below to illustrate how remarkable these materials were:

TYPICAL FORMULATION

Pluracol TP-440 triol	41.26	weight %
Teracol 2000 diol	17.57	weight %
Hylene W diisocyanate	41.17	weight %

DBTDL catalyst in drops per 100 grams of mixture to adjust work-life and so on.

PROPERTIES IN THE CURED STATE

It is a water-white transparent material suitable for optical applications.

This cast plastic had a dart impact resistance equivalent to that of polycarbonate.

It has superior resistance and marring resistance to that of polycarbonate.

However, it did poorly in accelerated weathering tests, even with UV absorbers and antioxidants in the formulation.

25.1.2   OUTSTANDING IMPACT RESISTANCE

Although the non-yellowing, water-white, transparent rigid polyurethanes never made an effective cladding material for polycarbonate windshield and canopies, the formulation effort generated several families of materials potentially useful for other applications. One of the things, which we learned is that we can formulate PU plastics, that are as impact resistant as polycarbonate plastic without the sensitivity to crazing from a host of common chemicals. That is quite an accomplishment because polycarbonate is astoundingly impact resistant. The weakness of these impact-resistant PU plastics is long-term outdoor weathering. However, there are hundreds of applications for high impact plastics which do not require long-term weathering stability.

25.1.3   EXCELLENT OUTDOOR DURABILITY

On the other hand, the harder, more cross-linked, PU plastics had excellent long-term weathering stability. They just did not have the exceptional impact resistance of polycarbonate. There are numerous applications for products having great outdoor stability and not requiring super good impact resistance. Outdoor signs are one example.

So what we have here is a continuous range of PU plastics to select from, great weathering, but lower impact resistance on one end of the range, and great impact resistance, but mediocre weathering resistance on the other end of the range. There is an infinite selection of intermediate PU plastics within the range.

25.1.4   UNRESOLVED PROBLEMS

One of the unresolved problems that we had when making castings of our non-yellowing, water-white, transparent rigid polyurethanes was finding a 100% effective release agent for casting it against glass surfaces. The polyurethane would adhere strongly to the glass even if there was only a pinpoint area free of release agent. Then a glass chip would be embedded in the PU plastics surface. The reason that we were casting against glass in the first place was to produce a very smooth optical surface. An opportunity for others is to investigate polished metal as a surface to cast against. Unlike glass, steel is too strong to chip and it may be a more difficult surface to adhere to.

25.1.5   SPIN-OFF POTENTIAL

The polycarbonate cladding application was difficult because so many different properties had to be improved at once. However, there are a host of other applications with simpler requirements. For example, the combination of water-white transparency, non-yellowing, and inherent impact resistance is adequate for making art objects, such as sculptures, jewelry, and so on. The harder, more weatherable formulations like Clad 5 would be suitable for casting lens and parabolic mirrors. The diversity of formulations that we explored makes customizing for new applications easier.

SHEET ADHESIVE

Some of our formulation variation studies examined the degree of cross-linking as a function of temperature. The least cross-linked formulations melted and flowed at temperatures above 240°F. These polyurethanes have great potential as hot melt adhesives. They could be used to laminate transparent materials such as glass, acrylic, polycarbonate, and so. For non-transparent applications, the hot melt adhesives could contain a filler which would reduce cost and improve thermal conductance of the adhesive, thereby speeding up the laminating process.

OTHER APPLICATIONS

1. Kits, packaged as a two-component casting compounds. Possible uses include molded transparencies such as motorcycles windshields, helmet shields, and so on. Custom cast art objects.
2. Transparent sheets: Sold in cast sheet form of various thicknesses. Shatter-proof transparency applications. Thermoformable sheet applications.
3. Encapsulations of electronics or objects needing an impact-resistant, see-through, plastic.
4. Cast jewelry applications requiring shatter-proof plastics and excellent color fastness. Bracelets, faceted beads, and so forth.

KEYWORDS

- specialty rigid thermosets
- polyurethanes
- polycarbonate

THE SCIENCE OF RUBBERY MATERIALS

CONTENTS

1.   GENERAL COMMENTS

There is a category of materials, which are soft, flexible, pliable, rubbery, or whatever adjective you would like to use. They are the opposite of hard and rigid. I want to be broad and inclusive in this category. The category spans the flexibility range from as soft as an artificial fishing worm to as hard as the rubber in a very firm tire. The category spans the range from rubbers that snap back instantly when stretched and released to rubbers that permanently deform when stretched. This property of resiliency brings to mind a story from my youth. My friends and I made our own slingshots using a "Y"-shaped branch and inner tube rubber. This was in the 1940s and WWII had forced the conversion from natural rubber to synthetic rubber. We could easily tell which was which because red inner tubes were natural rubber and black inner tubes were synthetic rubber. The red rubber made powerful slingshots whereas the black rubber was markedly inferior. I assume that either red or black rubber made functional inner tubes that held air pressure within a tire. Yet certain rubbers can be even much slower returning to their original shape than that black synthetic rubber, even taking several minutes. In order to understand why rubbers behave so strangely, we have to learn a little about the science of rubbery materials. I hope the following paragraphs help you see these intriguing materials in a new light.

Rubbery materials contain polymer chains that are constantly wriggling. Certain organic groups enhance free rotation, which accounts for the polymer chain flexibility. Ether and ester groups are examples. Also, the presence of a carbon-to-carbon double bond enhances free rotation to attached carbons. Rubbers also contain cross-links which prevent the chains from permanently slipping past each other. The distance of a chain segment from one tie point to the other is some multiple of the average distance due to the coiling and uncoiling. If we stretch a rubbery material, we can straighten the chain segments out. If we let go, the rubber seems to snap back to its previous state.

2.   UNUSUAL HEAT CHARACTERISTICS OF RUBBERY MATERIALS

When we say that we want to look at rubbery materials scientifically, a lot of math might be expected. If you enjoy partial differential equations and thermodynamics, try googling, "entropy and rubber." The math of thermodynamics is pervasive. On the other hand, I think most of my readers will get more out of these discussions if I avoid higher math entirely. I try to do just that in the discussions which follow:

2.1   ENTROPY

Elastomers behave differently than rigid materials with regards to heat. Heating causes them to contract, whereas cooling causes them to expand. If you have a wide rubber band handy, try this experiment. Put it to your lips and sense its temperature. Then stretch it quickly and put it to your lips again. You will sense that the rubber band is warmer. The harder and quicker you stretch it, the hotter it will get. The stretching process caused heat to be released. The explanation for this is that the polymer segments of a rubber are in constant motion. The usual length of the segment may be a fraction of the stretched out length. There are numerous different configurations that the wriggling segment may assume. In other words, there is a high state of randomness resulting from the freedom for segments to move within the molecular structure. Scientists have a name for randomness and that name is "entropy." Entropy is a term used in heat equations. The stretching drastically reduced the number of configurational possibilities and entropy of the rubber band was reduced. We sensed it as heat.

2.2   THE GOUGH–JOULE EFFECT

Consider a second experiment: Attach one end of a wide rubber band firmly to the edge of a table and attach weights to its other end. The rubber band will be stretched due to the weight. Carefully measure the length of the stretched rubber band. Next, bring a heat source close to the rubber band. A heat gun or hair dryer would work. Will the rubber band further elongate or will it contract? Answer: Heating the rubber band contracts it. It seems counterintuitive. If you have tested rigid plastics similarly, they always further elongate. Why is the rubber different? The answer is entropy. The heat added to the rubber band increases the curling up activity of chain segments, reducing volume of the elastomer and causing contraction. This is the force which pulls up on the weight.

This phenomenon of an elastomer contracting instead of stretching is known as the Gough–Joule effect. The effect was first witnessed by John

Gough in 1802. Fifty years later, James Joule looked into it in more detail, hence the name Gough–Joule effect. Note that if the elastomer had not been stretched out under tension, there would have been no contraction.

3. SNAP BACK IS AN ENTROPIC REACTION

When we stretched that rubber band it took work to do it, but when we released it, it snapped back on its own. What made it snap back? Remember that the rubber is a polymer network containing cross-links. The polymer chain segments are constantly wriggling so that the average distance between chain ends is a fraction of the stretched out distance. The cross-links anchor each chain segment end. When we stretch the rubber band, we continually reduce the number of different configurations for those chain segments. We are reducing the entropy of the rubber material. So, the force of a stretched rubber band is an entropic force. So when we release the stretched rubber band, it is entropy which causes it to snap back. Under tension, the polymer chains have been in a state of orderly alignment. When the tension is released, the polymer chains resume their random state of higher entropy. This is one of the unusual things about rubbery materials. The influence of entropy on the behavior of rigid plastics is so small that we ignore it.

4. MOLECULAR STRUCTURE OF THE POLYMER INFLUENCES FLEXIBILITY

Simply stated, polymeric materials are hard when the polymer chain is more inflexible and are soft when the polymer chain is more flexible. The chemical nature of the polymer chain components or of the side groups attached to it determine just how rigid or flexible the chain will be. For example, if I intend to design a polymeric molecule which has rigidity, the inclusion of benzene rings in the polymer chain or attached to the chain is a sure way to get there. Here is an example: epoxy plastics are hard and rigid because they are usually based on epoxy resins derived from bisphenol A, which has two benzene rings in its molecule. Other kinds of rings, such as cyclohexane, also add rigidity but not as much as the benzene ring does.

If we want to design a polymeric molecule which is rubbery, the absence of ring structures is helpful. However, the key to attaining flexibility is to increase free rotation around carbon-to-carbon bonds. Some of the ways of enhancing free rotation is through the addition of ether groups, ester groups, and carbon double bonds into the polymer chain. Hydrogen atoms attached to the carbons restrict free rotation because of their repulsive positive charge. When we reduce the number of hydrogen atoms as in the case of a carbon–carbon double bond, the single bond to the adjacent carbon atom has freer rotation and the polymer segment is more flexible.

Stereoregularity is another consideration in the design of polymeric molecules having flexibility. Stereoregular polymers have symmetry about an imaginary plane which bisects them. Such polymers can form crystallites due to their symmetry. Here is an example: Polyethylene is a stereoregular polymer and consequently the polyethylene plastic would be rubbery instead if it were not for the micro-crystals, which form within an amorphous polymeric matrix. These crystallites tend to stiffen the material. Polyethylene is essentially a two-phase material of crystallites and amorphous polymer and that is why it is translucent rather than transparent. So we can see that the stereoregularity in ingredients is an important thing to know. The formulating chemist can select either stereoregular or stereo-irregular ingredients depending on the property objectives. Although strength enhancement in elastomers occurs with stereoregular polymer chains, greater softness and transparency occur with stereo-irregular polymer chains.

The effect of cross-linking on properties is another variable. Natural rubber was of little use until the vulcanization process was discovered because the natural rubber would melt and become tacky on a hot day in its natural form. Vulcanization, which is just one particular method of creating cross-links, transformed natural rubber from a curiosity into a material with a thousand uses overnight. At the other extreme, too much cross-linking in an elastomer can render it useless. A cheesy condition develops in the excessively cross-linked elastomer. Consequently, it has very low tear resistance and very low extensibility. It may crumble in your fingers. The middle ground between these two extremes of tackiness and friability is the sweet zone, where useful elastomeric products can be created.

5.   ASSESSING THE THERMAL RANGE OF AN APPLICATION

We tend to identify a solid material by what it is at room temperature. It might be a hard plastic, a leathery material, a firm rubber, or a soft rubber. Yet, that same material may be quite different at a colder or warmer temperature. That hard plastic may behave as a rubbery material at a higher temperature and that soft rubber may be a hard plastic at a lower temperature. In either case, the material will pass through a unique temperature where it behaves as a leathery material. It will easily deform at that temperature and retract very slowly when the load is released. It is often described as being a "dead" material at this special temperature. Of course, that special temperature is known as the glass transition temperature.

The glass transition temperature is a fundamental property of a polymeric material and can tell us a great deal about how a given plastic or elastomer will behave in an application. We always want to know the temperature environment of an application. What is the highest temperature the material will see? The lowest temperature? The usual temperature? What will the material be expected to do at each of these temperatures? In general, we prefer to have the glass transition temperature fall outside the temperature range of the application. That way it is always an elastomer or a plastic in the service temperature range but not both.

I remember a commercial application, where a new customer was using one of our polyurethane elastomers for solid snowmobile tires. At colder temperatures, the tires would develop a flat spot on the bottom of the tire. The polyol in the polyurethane was a polyoxypropylene-type polymer. We replaced that polyol with a polyoxybutylene-type polymer and the problem vanished. It was simply a matter of lowering the glass transition temperature by an adequate amount.

6.   MEASURING TENSILE PROPERTIES OF ELASTOMERS

The methods for testing rubbery materials are so different from testing those same properties of plastics that rubbers have their own special test methods. Consider tensile testing, for example. Plastics are tested at a slow crosshead speed (0.05 inches/min) per ASTM D638 in order to provide a near-static test result. However, the ultimate elongation of rubbers might be as high as 400%. It would take far too long to get there at a slow crosshead speed. ASTM D412 standardizes on a crosshead speed of 20 inches/min. Most plastics have a Hookean behavior at least initially when stretched. It is possible to interpret a Young's modulus from the straight line portion of the chart. However, rubbery materials produce an entirely different kind of stress/strain curve. Instead of a Young's modulus, it is traditional to report a 100% tensile modulus. This is the stress measured at the point where the elongation is 100%. It is reported in units of psi (pounds per square inch).

INDEX